RED-FLESHED PEACHES

Copyright © C. Thornton 2015

The right of C. Thornton to be identified as author of this work has been asserted in accordance with the Copyright, Designs and Patents Act, 1988.

All rights reserved. No part of this book may be reproduced or transmitted by any person or entity (including Google, Amazon or similar organisations) in any form or by any means, electronic or mechanical, including photocopying, recording or by any information storage and retrieval system, without prior permission in writing from the publisher.

National Library of Australia Cataloguing-in-Publication entry

Creator: Thornton, C., author.

Title: Red-fleshed peaches / C. Thornton.

ISBN: 9781925110791 (paperback)

Series: Rare and heritage fruit cultivars ; set 1, no. 29.

Notes: Includes index.

Subjects: Peach--Heirloom varieties.
Peach--Varieties.
Peach.

Dewey Number: 583.73

ABN 67 099 575 078
PO Box 9113, Brighton, 3186, Victoria, Australia
www.leavesofgoldpress.com

RARE AND HERITAGE FRUIT
CULTIVARS #29

RED-FLESHED PEACHES

C. Thornton

- RARE AND HERITAGE FRUIT -
THE SERIES

SET #1

RARE AND HERITAGE FRUIT
- CULTIVARS -

1 Apples
2 Cider Apples
3 Crabapples
4 European Pears
5 Nashi Pears
6 Perry Pears
7 Apricots
8 Peaches
9 Nectarines
10 European Plums
11 Japanese Plums
12 Cherries
13 Figs
14 Cactus & Dragon Fruits
15 Oranges
16 Lemons

17 Limes
18 Mandarins & Grapefruit
19 Kumquats, Calamondins & Chinottos
20 Rare & Unusual Citrus
21 Nuts
22 Berries & Small Fruits
23 Quinces
24 Guavas & Feijoas
25 Table Grapes
26 Wine Grapes
27 Avocados
28 Rare & Unusual Fruits
29 Red-fleshed Peaches

and more...

SET #2

RARE AND HERITAGE FRUIT
- GROWING -

1 Propagating Fruit Plants (other than grafting)
2 Grafting and Budding Fruit Trees
3 Planting Fruit Trees and Shrubs
4 Care of Fruit Trees (compost, mulch, water etc)
5 Pruning Fruit Trees and Shrubs
6 Training and Espaliering Fruit Trees and Shrubs
7 Harvesting and Storage of Fruit
8 Pests and Diseases of Fruit Trees and Shrubs

SET #3

RARE AND HERITAGE FRUIT
- PRESERVING -

1 Fruit Preserving (drying, crystallizing, bottling etc.)
2 Cider Making
3 Perry Making ('pear cider')
4 Fruit Wine Making
5 Fruit Spirits and Liqueurs Making
6 Fruit Schnapps Making

www.leavesofgoldpress.com

CONTENTS

Introduction xi
About Rare and Heritage Fruit 1
About Peaches 11
Red-fleshed Peaches in France 23
Red-fleshed Peaches in New Zealand 31
Red-fleshed Peaches in Africa 37
Red-fleshed Peaches in the USA 39
Red-fleshed Peaches in Australia 49
About Pleaches 61
Red-fleshed Peach Recipes 63
    Desserts 64
    Iced Desserts 74
    Flans and Tarts 78
    Cakes and Slices 91
    Drinks 93
    Preserves, Conserves and Pickles 95
    Savoury Dishes and Salads 103
Red-fleshed Peach Liqueur 108
Index 112

INTRODUCTION

One of my passions is heritage (otherwise known as 'heirloom') and rare fruit. This interest was born from my love of words. Long ago I chanced to read, in a magazine picked up in some waiting-room or other, a fascinating list of names of old Cornish apple cultivars. 'Snell's Glass Apple', 'Pig's Nose', 'Pendragon'... who could resist?

My interest in the curious and enchanting names of heritage apples led to my obtaining some actual plants and growing them at the family farm on the Mornington Peninsula in Victoria, Australia. In turn, this led to my collecting other species of heirloom and unusual edible plants and joining groups dedicated to conserving these old varieties. Over time the collection expanded, filling up a walled garden, a greenhouse, several groves and garden patches and an orchard.

I realised, then, that the farm was at the perfect latitude, with the perfect local geology, topography and climate for growing a wide range of edibles. Peaches and apples, in particular, flourished in our orchard, thriving on the peninsula's rich volcanic soils.

With hundreds of heritage fruit cultivars to choose from I had to accept that I could not grow them all, and would be wise to specialise. This was easy. My greatest interest is in fruits and vegetables with unusual, richly-coloured flesh, such as purple-fleshed sweet potatoes and carrots, golden-fleshed tamarillos, cranberry-fleshed potatoes and guavas with crimson flesh. One kind of peach drew me like a magnet from the moment I first learned about them; the red-fleshed peaches.

I had never seen these unusual fruits. It was thanks to the Internet that I first learned of their existence. Until then, the only peaches I had ever seen possessed flesh that was either white or yellow.

Keen to grow these fruits, I spread the word amongst gardening friends and acquaintances that I was looking for some red-fleshed peach seeds or grafting material. At first I thought we would never find any in our part of the world. Gardeners in other countries appeared to have plenty of access to them, but nobody in Australia seemed to know anything about them.

It took two or three years of searching but eventually, through persistence and dedication, and with much help, I managed to hunt down some red-fleshed peach pits from various sources.

The lineage of these peach pits is a mystery. The people from whom I obtained them did not know anything of their history beyond the name of the person from who *they* had obtained them.

Fruit tree DNA often undergoes amazing journeys. For example, peaches are native to China, yet centuries ago they were carried all the way to Europe. After that, they made their way to the Americas, to Africa, and to New Zealand. Somehow, some even reached Australian shores.

We named the trees that grew from our precious peach-pits after the people from whose hands we received them. For example there is 'Greg's Red-Fleshed', 'Bob's Neighbour's Brother-in-Law's Red-Fleshed', and 'Susan's Red-Fleshed'.

Are our trees descended from the French Pêche de Vigne? Are they, perhaps, the progeny of New Zealand's Blackboy peaches, which themselves are probably heirs of the Pêche de Vigne lineage? Perhaps one day genetic technology will answer the question.

Each of our Mornington Peninsula red-fleshed peach trees has its own unique characteristics, but they all display the beautiful, solid, dark crimson-purple flesh and that delectable raspberry-tart flavour. They our our favourite peaches.

Naturally, as our trees grew bigger they produced more fruit year by year. When the trees were young they only bore sufficient fruit for us to eat, freeze and share with friends, neighbours and family. As they matured they became more and more productive, so that one year we found ourselves wondering what to do with the hundreds of kilos of fragrant fruit so bountifully betstowed upon us by our red-fleshed peach trees. We had a harvest glut!

There were simply too many to eat fresh, give away, sell at the local shops, freeze or preserve, so I did some research and discovered that a red fleshed peach liqueur could be made. In fact Pêche de Vigne liqueur is a popular drink in France. So, after some extensive experimentation, I ended up with a recipe for an amazing red-fleshed peach liqueur.

Having already written several books about other heritage fruit I thought it would be a good idea to put together all my red-fleshed peach research (and lots of recipes) and write this book.

C. Thornton
Mornington Peninsula,
Australia

ABOUT RARE AND HERITAGE FRUIT[1]

This book is one of a series written for 'backyard farmers' of the 21st century.

For the purpose of this series, rare fruits are species neither indigenous to nor commercially cultivated in any given region.

'Heritage' or 'heirloom' fruits such as old-fashioned varieties[2] of apple, quince, fig, plum, peach and pear are increasingly popular due to their diverse flavours, excellent nutritional qualities and other desirable characteristics.

It is much easier for modern supermarkets to offer only a limited range of fruit cultivars (i.e. varieties) to consumers, instead of dozens of different kinds of apples, pears etc. During the 19th and early 20th centuries, however, the diversity was huge. Old nursery catalogues were filled with numerous named varieties of fruits, nuts and berries, few of which are available these days.

What are heritage fruits? 'An heirloom plant, heirloom variety, heritage fruit (Australia), or (especially in the UK) heirloom vegetable is an old cultivar that is "still maintained

---

1   *Note: this introduction is identical in every handbook in the Rare and Heritage Fruit series.*
2   *The correct term in this case is 'cultivars'; however most people are more familiar with the term 'varieties' and although it is not strictly accurate, we use the terms interchangeably in this series.*

by gardeners and farmers particularly in isolated or ethnic communities".3

'These may have been commonly grown during earlier periods in human history, but are not used in modern large-scale agriculture. Many heirloom vegetables have kept their traits through open pollination, while fruit varieties such as apples have been propagated over the centuries through grafts and cuttings.'[4]

Broadly speaking, heritage fruits are historic cultivars; those which have initially been selected or bred by human beings and given officially recognised names, before being propagated by successive generations of growers, retaining their genetic integrity far beyond the normal life-span of an individual plant; those which are not protected by a private plant-breeders' licence, but instead belong to the public at large. They are the legacy of our ancestors; living heirlooms; part of humanity's horticultural, vintage and culinary heritage.

Fruit enthusiasts around the globe are currently reviving our horticultural legacy by renovating old orchards and identifying rare, historic fruit varieties. The goal is to make a much wider range of fruit trees available again to the home gardener.

This series of handbooks aims to help.

STORIES

Like people, every fruit cultivar has a name and a story. Take the Granny Smith apple, for example - the most successful Australian apple, instantly identifiable with its smooth green skin, exported world-wide, and now cultivated in numerous countries.

This famous cultivar began in the 1860s as a tiny seedling that chanced to spring up in a compost heap. An orchardist by the name of Mrs Maria Ann Smith lived with her ailing

---

3   Whealy, K. (1990). "Seed Savers Exchange: preserving our genetic heritage".(*Transactions of the Illinois State Horticultural Society* 123: 80–84.)

4   'Heirloom plants' Wikipedia. Accessed 2013

husband in Eastwood, New South Wales (now a suburb of Sydney). She was in her late sixties, a hard worker and the mother of many children.

One autumn day, as usual, Maria Smith drove her horse-drawn wagon home from the Sydney markets, where she had been selling the fruit from her orchard. The wagon possibly contained a few wooden crates she had purchased after selling her produce, in which to transport the next load of wares. One or two leftover Tasmanian-grown French Crab apples might still have been lying in the crates, somewhat battered and past their prime. Imagine 'Granny' Smith, her grey hair tucked up inside her bonnet, trudging down to the creek from which the household drew its water and dumping their decaying remains on its banks.

There in that damp spot, sinking into compost-rich soil, the apple pips lay throughout the winter months. Come spring, one of them split open and a tiny white rootlet appeared. It swiftly bored downwards, stood up and threw off its black seed-case, revealing two perfect, green cotyledons.

The leaves quickly multiplied as the seedling grew, Maria spied it next time she walked down to the creek, the hems of her long black skirts rustling through the ferns. She nurtured the infant tree until it grew up and bore fruit. When at last she picked the first green-skinned apple and took a bite, she must have been surprised by the crisp, hard flesh and sharp taste. No doubt she used it to make pies and other desserts for her sick husband and numerous grand-children, thus discovering that this new cultivar was good for both cooking and eating.

She shared the apples with friends and neighbours, allowing them to cut scion-wood from her tree and graft their own cloned versions. Locally, word of the apple's qualities spread.

'Smith died only a couple years after her discovery, but dozens of Granny Smith apple trees lived on in her neighbours' orchards. Her new cultivar did not receive widespread attention until, in 1890, it was exhibited as 'Smith's Seedling' at the Castle Hill Agricultural and Horticultural Show. The following year it won the prize for cooking apples under the name 'Granny Smith's Seedling'.

'The apple became a hit. In 1895 the New South Wales Department of Agriculture officially recognized the cultivar and began growing it at the Government Experimental Station in Bathurst, New South Wales, recommending its properties as a late-picking cooking apple for potential export.

'During the first half of the 20th century the government actively promoted the apple, leading to its widespread acceptance. However, its worldwide fame grew from the fact that it was such a good 'keeper'. Because of its excellent shelf life the Granny Smith could be transported over long distances in cold storage and in most seasons. Granny Smiths were exported in enormous quantities after the First World War, and by 1975 forty percent of Australia's apple crop was Granny Smiths. By this time the apple was being grown extensively elsewhere in the southern hemisphere, as well as in France, Great Britain and the United States.'

'The advent of the Granny Smith Apple is now celebrated annually in Eastwood with the Granny Smith Festival.[5]

NAMES

The origin of the Granny Smith apple is one of many intriguing fruit stories, but sometimes the name - or names - of cultivars tells yet another story, an etymological one. Names may be inspired by the place a new cultivar was discovered, by the person who selected or bred it, by the shape, flavour, colour or use of the fruit, by an event that took place around the time of discovery, by somebody's sweetheart, or any number of other factors.

Names, too, may be multiplied.

The Granny Smith apple was discovered after the advent of newspapers. If you forgot what the prize-winning cultivar was called, you could look it up and there it would be,

---

5   *'Granny Smith' Wikipedia. Accessed 2013*

in black and white. This was not the case for many ancient cultivars.

The Granny Smith apple's probable mother, the French Crab, itself boasts twenty-six listed synonyms, probably invented by forgetful apple-growers.

Another instance of numerous synonyms is the French cider apple whose name is Calville Rouge D'Hiver, meaning 'Calville Winter Red'. It arose in the late 1500s, and as its popularity spread across Europe, the first thing that happened was that people translated the name into their own language: 'Teli Piros Kalvil', 'Roter Winter Calville, 'Calvilla Rossa di Pasqua', 'Cerveny Zimni Hranac' etc.

Next, when absent-minded peasants could not remember the name of this excellent red fruit, they gave it another one. Imagine a weather-beaten farmer in some isolated French village scratching his beard and musing, 'It was something to do with "Calville". 'Calville Rouge,' perchance?' Across the valley in another village, a cider-brewer was knitting his (or her) puzzled brow and saying, 'It was something to do with winter, I am thinking, or was it autumn? "Pomme d'Automne"?' Further afield, a third Frenchman shrugged his shoulders and declared, 'Devil take me if I can remember how it is called, but it is big and red like the heart of a bull, so let us name it "Coeur de Boeuf."'

Fanciful, perhaps, but this might explain why, on the database of the UK's National Fruit Collection, there are more than a hundred synonyms listed for Calville Rouge D'Hiver.

Words are forever evolving. Even when cultivar names stay the same, the language around them is changing and their original meaning becomes lost in the mists of time.

One example of this is the grape cultivar Cabernet Sauvignon, which is considered a relatively new variety, being the product of a chance 17th century crossing between Cabernet franc and Sauvignon blanc.

'Cabernet franc' can be etymologically traced back to 'French Black Grape' (from the Latin word 'caput' which means 'black vine'). The word 'Sauvignon' is believed to be derived from the French 'sauvage', meaning 'wild' and to refer to the grape being a wild grapevine native to France. 'Blanc,' of course, means 'white'. 'Cabernet Sauvignon' no longer means 'Wild Black Grape' in modern French - that would translate as something like 'Vigne Noir Sauvage'. The ancient cultivar name has now taken on its own meaning and is virtually synonymous with the wine made from it.

It is interesting to compare typical cider apple names with, say, typical peach or perry pear names. French words abound among heritage cider apple cultivars, reflecting their roots in medieval Normandy. To the ears of English-speakers these names may sound rather mysterious and aristocratic, until you translate them: for example, Gros Bois, Jaune de Vitré, Moulin à Vent du Calvados, Noël des Champs, Belle Fille de la Manche, Petite Sorte du Parc Dufour and Groin D'âne translate respectively as Big Wood, Yellow Glass, Windmill of Calvados, Christmas Field, Beautiful Girl of the English Channel, Small Kind of Park of the Oven and Donkey's Groin.

Some names of heritage perry pears give us an insight into the bawdy, rustic humour of the perry-drinking English peasants who originally selected them; Ram's Cods, Startle Cock and Bloody Bastard to mention a few.

Heritage grape cultivars have names that come from all over Europe, particularly France and Italy.

Figs go back even further. Humans were cultivating them around 9400 BC, a thousand years before wheat and rye were domesticated. Their names, in English at least, are often drawn from their colour and their place of origin - Brown Turkey, White Adriatic, Black Genoa, Pink Jerusalem, Green Ischia ...

Peaches, a more 'modern' fruit in terms of their popularity and breeding, often bear invented names with fancy spellings, such as Florda Glo, Earligrande, Harbrite and Dixigem.

'IMMORTAL' DNA

Another major difference between stone fruit and fruits such as grapes, figs and apples is their ability to grow 'true' to their parents from seed. Stone fruits are far more homozygous than their ancient cousins the pomes (apples, pears etc.) and the grapes. Growers do graft them, but if you plant their seeds the new tree will bear fruit that's fairly similar to that of the parent tree. This means that the centuries-old grafting traditions, the fierce cherishing, the careful bequeathing and the meticulous labelling that accompany pome fruits, grapes and other heterozygotes are not seen as often in the world of peaches and nectarines. This is why many of their cultivar names seem so different, arising as they do from highly organised commercial breeding programmes of the 20th and 21st centuries.

Unlike the seedlings of say, peaches and nectarines, seedling apples are an example of 'extreme heterozygotes', in that rather than inheriting DNA from their parents to create a new apple with those characteristics, they are instead significantly different from their parents.'6 (Humans are rather like apples in that way, though not as extreme.)

Returning to our green-skinned Australian apple - 'Because the Granny Smith is a chance (and rare) mutation, its seeds tend to produce trees whose fruit have a much less appealing taste. To preserve the exact genetic code of any plant variety, a stick of the wood has to be 'cloned'. It has to be grafted onto new roots (or planted directly into the ground, but this is uncommon for trees). Thus, all the Granny Smith apple trees grown today are cuttings of cuttings of cuttings from the original Smith tree in Sydney.'7

---

6   *John Lloyd and John Mitchinson (2006). QI: The Complete First Series – QI Factoids*
7   *Stirzaker, Richard (2010). Out of the Scientist's Garden: A Story of Water and Food. Collingwood, VIC: CSIRO Pub.*

Cloning by grafting means that the heritage trees - and shrubs - which have survived through the years are genetically identical to their ancestors. Indeed, the heritage plants of today possess exactly the same genetic code as the original trees that arose centuries ago in Asia and Europe. For example, another heritage apple cultivar, 'Court Pendu Plat', is thought to be 1500 years old - the oldest one in existence. Introduced into Europe during Roman times, the living wood from that same tree flourishes to this day, right here in the Great Southern Land.

COMMERCIAL CULTIVARS

Naturally, plant breeders strive to provide the products demanded by the market. Commercial orchardists want to purchase heavy-bearing trees with high disease resistance, whose fruit ripens all at the same time to save on picking costs. Wholesalers want fruit that keeps in storage for a long time without spoiling, and can be shipped without damage. Only firm-fleshed, bruise-resistant fruit will survive modern-day processing. After harvesting, apples, for instance, are tipped into crates, then passed along a conveyor belt through machinery that washes and brushes them clean of insecticides and dirt. This process removes some of the fruit's natural protective coating, so the machines re-apply a commercial grade wax before polishing them to a high shine and pasting a plastic label onto each one. Then the apples are packed into cartons for shipping to markets and stores.

Supermarket shoppers demand visually attractive fruit - large, regular in shape, unblemished and with highly coloured skin. Consumers also choose fruit with extra sugar content and juiciness.

All these characteristics, nonetheless, do not necessarily give rise to the best flavour or nutrition. To pick a tree-ripened fruit from your own back yard and bite into it is to experience the taste of fresh food as our forefathers knew it. Growing and preserving their own food, unconcerned with transportability and long storage times, they aimed for a wide variety of fruits, each of which had a unique and delicious taste.

Rare fruit, heritage and heirloom fruit enthusiasts across the world are reviving our horticultural legacy by renovating old orchards and sourcing 'lost' historic and unusual fruit varieties. Their goal is to encourage community participation and to make a wide range of fruit trees available again to the home gardener.

This series of handbooks aims to help.

WHY PRESERVE RARE AND HERITAGE FRUITS?

- They provide access to a wider range of unique and delicious flavours.
- We can enjoy the nutritional benefits of fresh, tree-ripened food.
- Biodiversity: The preservation of a wide range of vital genetic material helps to insure against the ravages of pests and diseases in the future.
- They allow a longer harvesting season, with early and late ripening.
- Culture: heritage varieties, with their interesting assortment of names, are living history.

**Collections of heritage fruit trees are precious.
Anyone who is the custodian of an old tree should treasure it.**

ABOUT PEACHES

HISTORY

Peach trees are native to northwest China, which is where mankind first cultivated them. Old manuscripts indicate that peach cultivation in that country dates back to 1100 BC. China is still the world's largest producer of peaches.

The original, wild peaches were small, sour and very fuzzy-skinned. Over the centuries, people chose to plant seeds from the sweetest and largest fruits, leading, over thousands of generations, to the peaches we know today.

Peaches are not only a popular fruit, but are symbolic in many cultural traditions, such as in art, paintings and folk tales.

Over the years travellers carried peach seeds—and possibly small potted or bagged trees, too—from China to Persia (now Iran), where they were widely cultivated for their delicious, juicy fruits. From Persia, peaches were eventually brought into Europe.

These fruits were well known to the Romans in first century AD. U.P. Hedrick in his book 'The Peaches of New York'[1] wrote: 'Peach-culture in France probably began as early as in Italy, for both Columella and Pliny mention the peaches of Gaul with those of Rome. Introduced thus early, finding

---

[1] *Albany J. B. Lyon Company, Printers. 1912*

suitable soil and climate and easily propagated, so delicious a fruit as the peach must at once have become a prime favorite in the orchards of the monasteries, where, tended by monks who were the most skilled horticulturists of the times, the peach was disseminated throughout France with the spread of Christianity. France was the foster-mother of the peach in Europe—from her nurseries the Belgians, Dutch, Germans and English bought their first peach-trees. The history of the peach in France, then, is an important chapter in the history of this fruit."

European explorers brought peaches to the Americas. Now they are cultivated all over the world.

Scientific Classification

The scientific classification of peaches is family: Rosaceae; genus: Prunus; subgenus: Amygdalus, species: persica.

At the time when the great 18th century botanist Linnaeus was devising a botanical name for peaches, it was believed the fruit originated in Persia, which is why he named the variety 'persica' — from the Latin 'Persae' meaning 'Persians'.

The term 'prunus' comes from the Latin word for 'plum tree'. The genus Prunus includes plums, cherries, peaches, nectarines, apricots and almonds.

The peach is classified with the almond in the subgenus 'Amygdalus', because the kernels of both of them are almond-shaped.[2]

---

2   *Interestingly, there are parts of our brains whose names derive from their similarity in shape to these kernels—the amygdalae. The amygdalae play an important part in the processing of our memories, decisions and emotions.*

Differences between Peach Cultivars

Some of the following information on the different types of peaches has been drawn from my book 'Peaches: Rare and Heritage Fruit Cultivars #8'.

Flesh Colour

Peaches may be either yellow-fleshed, white-fleshed or red-fleshed. Studies have shown that in general the western palate prefers yellow- or red-fleshed peaches while Asian consumers prefer white flesh.

Because white peaches have lower acid levels and higher sugar levels, they also have a slightly sweeter taste. White peaches ripen faster, are more fragile and bruise more easily; thus they are harder to transport.

Red-fleshed peaches are rare in some countries, though they are popular in France (where they are called 'Pêche de Vigne', 'pêche sanguine' or 'pêche sanguine vineuse') and New Zealand (where they usually go by the name of 'Blackboy'). In the USA they go by the name of 'Indian Blood' and in Australia they are often simply called 'Blood' peaches. Aside from their ruby-coloured flesh one characteristic of red-fleshed peaches by any name appears to be that they are propagated from seed rather than by grafting.

A similar red-fleshed fruit is the pleach, which is not a true peach but a cross between peach and a blood plum.

The 'Pit' or 'Stone'

Peaches and nectarines are either clingstone, freestone or semi clingstone. Inside the hard pit is the nut-like kernel.

Clingstone—The flesh adheres to the pit when the fruit is cut in half. Most early-season peach and nectarine cultivars are clingstones. Clingstone cultivars tend to have firm flesh.

Freestone—Cultivars with flesh that separates easily from the pit. Most cultivars used for fresh eating are freestone.

Semiclingstone or semifreestone—Fruit whose flesh separates easily from the stone when the peach is fully ripe.

The red-fleshed Blackboy peach is freestone. The French Pêche de Vigne can be either clingstone or freestone, and in the USA the red-fleshed clingstone peach is known as 'Cherokee' or 'Indian Blood Cling' while the freestone version is called 'Indian Free'.

Flesh Texture[3]

Peaches are classified by flesh texture as either melting, nonmelting or hard.

Melting flesh peaches become softer as they ripen and will 'melt in your mouth' when they are fully mature. Most consumers prefer this type for fresh eating out of hand.

Nonmelting flesh peaches remain firm in texture when fully mature and never become melting Nonmelting flesh peaches typify most peaches that are used for commercial canning because they need to keep their shape. Some 'freestone, melting' types are canned but they represent a very small proportion of canned peaches.

The hard flesh type is very firm, even crispy when fully ripe. This type never melts and is typical of some white fleshed peaches from Asia.

Uses

Red-fleshed peaches may be eaten raw, cooked, sweet, salted, flambée, iced or deglazed. They also combine with wine, olive oil, basil, tarragon, mint, cardamom or ginger, white pepper, and hawthorn blossoms. Red-fleshed peaches are used in in cake recipes, compotes and jams.

Most peaches are classified as 'dessert' fruit because they are delicious eaten fresh no matter what type of pit they have. Some, however, are better suited for culinary purposes.

The semi-free and freestone are the best peaches for making pies, freezing, drying and canning etc. because the flesh of these peaches is easier to separate from the pit.

---

3   Source: Clemson Cooperative Extension

The semi-freestone and clingstone can be used for all purposes, but they require a little more work than the freestone fruits. Many people prefer to use the clingstone cultivars for jams, preserves, jellies, and pickles, as their flesh is firmer.

During a normal season, clingstone peaches ripen first, followed by semi-free, and freestone peaches in the order they are listed. Keep in mind that each season depends on the weather, and if there is a warm spring, peaches usually ripen earlier. After a cool spring, they generally ripen later.

Peach Blossoms

Peach blossoms are highly ornamental. In colour they range from light pink, to carmine, to almost purple. Their beauty is so highly valued in countries such as China, Japan and the USA that annual Peach Blossom Festivals are held in celebration.

The flowers of the peach tree can also be eaten raw or cooked. They are sometimes added to salads or used as a garnish. They can also be brewed into a tea. The distilled flowers yield a white liquid which can be used to confer a nutty, almond-like flavour like that of the seed. The seed, too, can be eaten raw or cooked, however this is not advisable. The seed can contain high concentrations of poisonous hydrocyanic acid.

In Chinese culture peach blossoms are highly prized The ancient Chinese believed that peach trees possessed more life-giving properties than other trees because in spring their blossoms appear before their leaves.

Nectarines

Most peaches have fuzzy skin. Smooth-skinned peaches are called nectarines. They originated as mutations of Prunus persica, and are known as 'Prunus persica var. nectarina'.

In English speaking countries, all the smooth skin peaches are named "nectarines".

True-to-Type

The majority of common fruit trees, such as most apples, require a pollinating partner (a partner tree of the same species). With cross-pollination, pollen is transferred from one tree to another by an agent such as an insect, or by the wind. The pollen fertilises the flowers, and the tree bears fruit.

When a fruit has two parent trees, it contains a mixture of genetic material from both of them. This means the seeds in that fruit may grow into trees with characteristics that are very different from the mother tree.

Nearly all peaches are self-fertile (also called self-pollinating or self-fruitful), meaning that one peach tree all by itself can produce fruit. The blossoms of self-fertile trees do not need to be fertilised by pollen from another tree to bear fruit. With self-pollination, a blossom's stamen drops pollen directly onto its own stigma (type 1 pollination) or pollen is transferred from the anther of one flower to the stigma of another flower from the same plant (type 2 pollination).

When peach trees use their own pollen, the new generation of peach fruits and seeds does not acquire new genetic traits. Without new genetic input, they become replicas of their mother tree.

> "The peach is a tree which is best grafted on almond or plum stock grown from seed; it is a member of the Rosaceae family. But one variety of clingstone peach, the Pêche de Vigne, reproduces true to type."[4]

When seeds become trees that closely resemble the mother tree, we say they are 'true-to-type'. Seedling peach trees usually grow relatively true-to-type.

---

4   From "A History of Food," by Maguelonne Toussaint-Samat. Publisher: Wiley-Blackwell; 2nd edition (November 3, 2008)

However, even self-fruitful trees bear more fruit if they are cross-pollinated by another tree, and all peach trees are capable of receiving pollen from each other.

Grafted fruit trees of most varieties tend to have shorter lifespans than seedling trees. Red-fleshed peach trees, since they are propagated from seed, live longer than grafted peach trees.

Genetic variations do happen. Peach trees may receive pollen from other peach trees with a slightly different genetic makeup, and new genes may arise via spontaneous mutations in existing genes.

In each succeeding generation of peach trees, even when grown from seed, there may arise tiny differences in various characteristics such as size, shape, colour, flavour etc. produced by natural mutations. Over time, these discrepancies diverge further and further apart.

This might explain the differences between peaches from different countries. For example the Blackboy peaches of New Zealand and the Indian Blood peaches of the USA are disease-resistant and hardy, while the Pêches de Vigne from France are prone to mildew. The Indian Blood peaches may be clingstone or freestone, while the Blackboys are all freestone. There are differences, too, in sweetness, acidity, flavour, flesh colour, skin colour, flesh density, juiciness, ripening times etc.

Even more differences can be induced by variations in terroir.

Terroir

Peach trees can only flourish and be productive in certain climates. Most peach cultivars need to experience very low temperatures for a certain length of time in winter, before they will produce any fruit. This is called their 'chilling requirement'.

The chilling requirement of a fruit is the minimum period of cold weather after which a fruit-bearing tree will blossom. It is often expressed in chill hours, which can be calculated in different ways, all of which essentially involve adding up the total amount of time in a winter spent at certain temperatures. Most peach cultivars require between 600 and 1,000 hours of chilling. Low altitude tropical regions cannot supply this chilling requirement. In tropical and subtropical latitudes, peach trees may grow at higher altitudes where it is cooler.

Peach blossom appears early in spring. This means that in cooler climates, any late frosts can cause the flowers to shrivel and die.

Peach trees like to grow in light, open, well-drained soil that can hold moisture. They dislike heavy, wet ground. Their preferred soil pH is neutral to slightly acid, but if the pH drops below 6.5 the acid conditions will prevent the shell of the peach kernel from hardening. As a consequence, the centre of the fruit decays.

Unlike apples, peaches do not require cool nights to develop red skin colour. Any red colouring on their skin depends more on the particular cultivar and how much light the tree receives.

Nectarines are even less hardy than peaches. Both nectarines and peaches grow best in warm, sunny, sheltered positions.

Peach trees will generally only be productive for about 9 years. Typical peach cultivars begin bearing fruit in their third year and have a lifespan of about 12 years.

Even if peach trees are given the right soil pH and fertility, the right number of chilling hours and shelter from late frosts at blossom time, they still have to experience hot summers before their fruit crop can ripen. Summer heat is required to mature the crop, with mean temperatures of the hottest month between 20 and 30 °C (68 and 86 °F).

Peaches are frequently cultivated in 'Mediterranean' climates. This type of climate is characterized by hot, dry summers and cool, wet winters. 'Mediterranean' climates are located between about 30° and 45° latitude north and south of the Equator.

Australia's Mornington Peninsula in Victoria is ideally located for peach growing, at 37.8° south of the equator. Its maritime influence contributes to its unique 'terroir'.

In the USA California is the leading peach producing state. Its latitude is about 37° north of the equator. California is also the largest and most important wine region in the United States. With mountains, valleys, plains and plateaux, the state's topography is as complex as its climate, offering winegrowers a bewildering choice of terroir.

Commercial peach production in China can be classified into seven regions based on terroir. In each of these regions, widely differing peach cultivars are grown. It is in the Northern China Plain region—the most important peach production region in China—that red-fleshed peaches flourish. The five top provinces for peach production are Shangdong, Hebei, Henan, Hubei and Jiangsu. Shangdong Province produces a red-fleshed peach called 'Red Gem'.

At 45.67° north of the equator, the French town of Soucieu-en-Jarrest lies on the very fringe of the perfect 'Mediterranean' latitude range. Soucieu-en-Jarrest is near the city of Lyon, in the south of France, in the 'Lyonnaise' region.

'Lyon lies in the broad transition zone between the Temperate Oceanic climates of northern France, and the Mediterranean climates to the south. Although Lyon does not share the drier summers typical of Mediterranean climates, it has summer temperatures that are warmer than typical temperate oceanic climates.[5]

---

5   *Wikipedia 'Lyon'. Retrieved 25th March 2015*

Red-Fleshed Peaches

Red-fleshed peaches are far rarer than the white-fleshed or yellow-fleshed types. In some countries they are almost impossible to find. How they spread from China to other parts of the world is an interesting story, partially shrouded in mystery.

Their colour sets these fruits apart — the deep ruby shade of their flesh make them spectacular additions to recipes. Their flavour, too, is unique. Moreover, red-fleshed peaches have numerous health benefits. Rich in antioxidant anthocyanins and flavonoids, they possess qualities that both heal and protect the human body.

Health Benefits of Red-Fleshed Peaches

'Anthocyanins are water-soluble pigments that may appear red, purple, or blue depending on the pH. In flowers, bright reds and purples are adaptive for attracting pollinators. In fruits, the colorful skins also attract the attention of animals, which may eat the fruits and disperse the seeds. In addition to their role as colour-givers, anthocyanins also act as powerful antioxidants.'[6]

Anthocyanins are abundant in certain deeply-coloured fruits — for example berries in the Rubus genus (raspberries, blackberries etc.), berries in the Vaccinium genus, (blueberries, cranberries etc.) berries in the Ribes genus such as blackcurrant, Prunus fruit such as red-fleshed peaches and cherries, and dark-coloured grapes (Vitis genus). Anthocyanins are also plentiful in eggplant peel, black rice, red cabbage, and violet petals, which are edible.

'Flavonoids are the most important plant pigments for flower coloration, producing yellow or red/blue pigmentation in petals designed to attract pollinator animals.

---

6 *Anthocyanin. From Wikipedia, the free encyclopedia. Retrieved 01.04.15*

'Flavonoids have been shown to have a wide range of biological and pharmacological activities in in-vitro studies. Examples include anti-allergic, anti-inflammatory, antioxidant, anti-microbial (antibacterial, antifungal, and antiviral), anti-cancer, and anti-diarrheal activities.'[7]

Red-fleshed peaches are rich in health-giving compounds. Red colour in fruit is important not only as an indicator of ripening, but also for its health benefits in the form of flavonoids and anthocyanins, which are the main components of red pigments in fruits, and possess antioxidant activity.[8]

In Japan, for example, because most consumers peel apples before eating them, scientists have been engaged in a project for the breeding of new apple cultivars with red flesh.[9] The idea is that red-fleshed apples should prove attractive to consumers, given their novel color and health benefits.

'Anthocyanin pigments and associated flavonoids have demonstrated ability to protect against a myriad of human diseases.

'Anthocyanins are members of the flavonoid group of phytochemicals, a group predominant in teas, honey, wines, fruits, vegetables, nuts, olive oil, cocoa, and cereals. The flavonoids ... comprise a group of over 4000 C15 aromatic plant compounds. The colorful anthocyanins are the most recognized, visible members of the bioflavonoid phytochemicals.'[10]

Anthocyanins and anthocyanin-rich mixtures of bioflavonoids may provide protection from genetic mutation and the development of hormone-dependent diseases. They may regulate immune responses, have anti-inflammatory activity,

---

7   *Flavonoid. From Wikipedia, the free encyclopedia. Retrieved 01.04.15*
8   *Eberhardt et al., 2000; Wolfe et al., 2003*
9   *Efficient Breeding System for Red-fleshed Apple Based on Linkage with S3-RNase Allele in 'Pink Pearl' . Keiko Sekido, et.al. HortScience April 2010 vol. 45 no. 4 534-537*
10 *Anthocyanins and Human Health: An In Vitro Investigative Approach. Mary Ann Lila. J Biomed Biotechnol. 2004 Dec 1; 2004(5): 306–313. doi: 10.1155/S111072430440401X PMCID: PMC1082894*

decreasing the permeability and fragility of blood vessels and strengthen membranes in the body.

'The roles of anthocyanin pigments as medicinal agents have been well-accepted dogma in folk medicine throughout the world, and, in fact, these pigments are linked to an amazingly broad-based range of health benefits.

'For example, visual acuity (eyesight) can be markedly improved through administration of anthocyanin pigments to animal and human subjects, and the role of these pigments in enhancing night vision or overall vision has been particularly well documented.

'In research trials, anthocyanins have demonstrated marked ability to reduce cancer cell proliferation and to inhibit tumor formation.

'The role of anthocyanins in cardiovascular disease protection is strongly linked to oxidative stress protection.'[11]

Tsuda et al[12] showed that anthocyanins extracted from purple corn, when provided to mice who were eating a high-fat diet, effectively stopped them from gaining fat. Symptoms of high blood sugar, excess insulin and hyperleptinemia (the presence of a higher than normal amount of leptins in the bloodstream) caused by the high-fat diet did not occur when mice also ate anthocyanins. The experiments suggested that anthocyanins in food can help prevent obesity and diabetes.

Anthocyanins are thought to be able to modulate cognitive (thought) and motor (movement) function, to enhance memory, and to help prevent age-related deterioration in brain function. Lab studies show that anthocyanins exert several protective effects against pleurisy.[13]

---

11   Ibid.
12   Tsuda T, Shiga K, Ohshima K, Kawakishi S, Osawa T. *Inhibition of lipid peroxidation and the active oxygen radical scavenging effect of anthocyanin pigments isolated from Phaseolus vulgaris L. Biochem Pharmacol. 1996;52(7):1033–1039. [PubMed]*
13   *Anthocyanins and Human Health: An In Vitro Investigative Approach. Mary Ann Lila. J Biomed Biotechnol. 2004 Dec 1; 2004(5): 306–313. doi: 10.1155/S111072430440401X PMCID: PMC1082894*

Red-Fleshed Peaches in France

The History of Peaches in France

Following the conquest of Persia, Alexander the Great introduced peaches into Europe under the name 'pecta'. In France the Latin term for peach, *malum persicum*, (Persian apple) evolved into 'pesche' (12th century), and by 1740 it had become 'pêche'.

Peaches have been grown in France since the sixth century. By the time Francis I took the throne in 1559, France boasted at least forty peach varieties.

Many peaches were grown in Île de France, the region surrounding Paris and Versailles which in modern times is covered with suburbs. Peaches were grown close to the markets of Paris because these fragile fruits did not stand up well to transportation by horse and cart — they bruised too easily. Between the sixteenth and nineteenth centuries, significant production of peaches in espaliered form took place along the walls of Montreuil-sous-Bois. The town, just outside Paris, became known as 'Montreuil-aux-Pêches'.

The climate of Île de France is cooler than the peach-growing regions of the south. Training trees to spread their boughs flat against masonry walls was necessary in order to get peach trees to bear fruit in these higher latitudes. It was the Chinese who invented the art of espalier and the French who gave the system its modern name. Peach growers used to espalier the peach trees so that the fruit would catch as much sun as possible. The masonry walls would absorb the sun's heat during the day and give it off at night, thus warming the peaches. Despite the attentive care of growers, peaches in Île de France did not ripen until the end of the summer, two months later than in the south.

By the seventeenth century peaches had become very popular with the French aristocracy Peaches were one of the favorite fruits of King Louis XIV, the 'Sun King': he grew thirty-three different varieties in his 'potager' (kitchen gardens) at Versailles, under the care of his gardener Jean-Baptiste de La Quintinie.

De la Quintinie was able to make peaches ripen even in the cooler climate of Versailles. He introduced a system of espaliering the trees against white-washed masonry walls that faced south. The whiteness would reflect sunlight onto the peaches and the masonry would store the sun's heat to warm them after sunset. This system proved successful and Louis was able to enjoy fresh peaches.

De la Quintinie also insisted that espaliered fruit trees must be beautiful to look at from every angle. Trained as a lawyer, this food garden enthusiast established a reputation for designing and managing the most extraordinary kitchen gardens that the people of Paris had ever known.

The Sun-King's Head Gardener wrote a gardening book called 'Le parfait jardinier, ou, instruction pour les jardins fruitiers et potagers' (The perfect gardener, or, instruction for fruit and vegetable gardens). The book contains a diagram showing a garden layout. One section of the garden is devoted to 'peches violette', which translates as 'purple peaches'. Are these, perhaps, the red-fleshed type?

It was Jean-Baptiste de la Quintinie who gave peaches an important place in the king's gardens among the fruits of September. He wrote that September was the real month of good peaches. They could be 'piled up like great pyramids at every meal; each variety ripening in the order of maturity that nature had set for them, thus providing them copiously and successively during the whole month.'

For almost a century the French court had enjoyed a culture of liqueur drinking—at least since Catherine De Medici wed Henry II of France and introduced liqueurs from Italy. Peach liqueurs were among the favourites of royalty and the aristocracy.

From the nineteenth century, peaches became the foundation of elaborate French desserts. Records show that the granddaughter of the Duchess of Berry preferred them with caramel. In 1829 the pastry chef Marie-Antoine Carême created a nectarine iced punch for Rothschild. In 1899 the pastry chef Auguste Escoffier invented the Peach Melba (peaches, vanilla ice cream and raspberry puree) in honour of the Australian singer Nellie Melba.

French Terms for Peaches

In French shops and supermarkets peaches—that is, the fruits with downy skins—are called 'les pêches'[14]. This holds true whether they are clingstone or freestone.

However French nurserymen, fruit connoissors, scientists, botanists and collectors called freestone peaches 'pêche vraie'[15] (true peach), and clingstone peaches 'pavie'.

In general, 'pêche vraie' cultivars have soft flesh while 'pavie' cultivars have firm flesh.

The French classify nectarines as either 'nectarine', which refers to freestone nectarines, or 'brugnon', which refers to clingstone nectarines. Nectarines and brugnons can have white or yellow flesh — the only difference between them is the type of pit.

In the 21st century a French plant breeder called René Monteux-Caillet developed a 'nectarine' (French meaning) with red flesh, under the brand name 'Nectavigne'®.

---

14   Confusingly, the French word for ';fishing' is identical to the French word for 'peach' — that is, 'pêche'.

15   The terms 'pêche vulgaire' and 'pêche proprement dite' are sometimes used as synonyms for 'pêche vraie', and the word 'persèque' is a synonym for 'pavie'.

Pêche de Vigne

A favorite late summer treat in France is the Pêche de Vigne. This term translates literally as 'Peach of the Vine', but it can also be translated as 'Peach of the Grapevine' or 'Vineyard Peach'. Pêche de Vigne is not a variety but rather a type of peach with many varieties of its own. It flowers late, and its fruits ripen late. Some of the Pêche de Vigne varieties have white flesh or yellow flesh, but many are different from other peaches in that their flesh is red.

Somewhat bewilderingly the French also use the term 'Pêche de Vigne' to describe the fruit from trees whose shape is 'en plein-vent' ('exposed to the wind')— that is, trees that are free-growing and have not been espaliered.

Pêche de Vigne trees traditionally play an important role in French wine production, particularly in the south, and their fruits are so highly regarded that an annual festival is held in their honour.

In France's Bordeaux region, the official source of Bordeaux wines, the vignerons (wine-grape growers) plant roses at the head of every row of grapevines. The roses look beautiful in bloom, however their purpose is not to be ornamental, but to be useful. Being susceptible to mildew, they will show signs of a mildew attack before the grape vines do, thus enabling the vignerons to start treating the vines while the disease is still in its early stages. This gives a better outcome.

Due to the fact that Pêche de Vigne fruits ripen at around the same time as the Lyonnaise grapes, the Coteaux du Lyonnais vignerons plant Pêche de Vigne trees at the ends of the vine rows, using them as indicator plants in the same way that roses are used in the Bordeaux region. Both the rose and the Pêche de Vigne are prone to mildew, so they give early warning of the disease. In France, where every region has its specialty foods, the red-fleshed Pêche de Vigne is considered a specialty unique to the Lyonnaise region.

The common feature of Pêches de Vigne is that their fruit ripens in the same period as grapevines. The grape-gatherers, at wine-harvest time, pick the peaches and enjoy them at the end of their day's work.

That is how the Pêche de Vigne got its name.[16]

A Pêche de Vigne enthusiast and collector in Saint-Etienne d'Estrechoux in the Herault has travelled the length and breadth of France collecting as many varieties as possible. He grows 120 different named varieties in his collection, whose fruits ripen over a period of five months. Many of them have red flesh.

The Red-fleshed Pêche de Vigne

The skin of the red-fleshed Pêche de Vigne is fuzzy and dusky in colour —a kind of warm, pnkish-grey. The flesh within is a dazzling crimson; the colour of rubies and blood. These sumptuous fruits ripen over a short period, during the final weeks of summer. In the southern hemisphere their harvest season lasts from early February to mid-March; in the northern hemisphere they are usually ripe from August to mid-September.

Popular fruits always accumulate synonyms. The wine-red colouring of the French Pêche de Vigne earned it several names. It is also known as 'Peche Vineuse' ('Vinous Peach'), 'Peche Sanguine' ('Blood Peach') and 'Pêche Sanguine Vineuse'.

One of the differences between a standard peach and a Pêche de Vigne is that the latter are propagated not by grafting but from seed.

A favourite treat in France, their taste is an intriguing blend of ripe peaches and juicy berries. They are more flavourful than normal peaches, but less sweet, and sometimes even slightly tart. They are not sugary-sweet but tart, like raspberries. Their

---

16   *In Switzerland, the name 'Pêche de Vigne' refers to a late-ripening, white-fleshed peach with a fuzzy red and green skin. The Swiss call the red-fleshed types 'Cardinal' peaches.*

flavour hints of plum, peach, Morello cherry and the richest red wines.

Some call the red-fleshed Pêche de Vigne 'the most glorious peach on earth', and 'the most delicious peach there is.' It has been described as having a 'deep raspberry-rose skin and flesh stained a deep red all the way through. Deliciously sweet and intensely perfumed.' (From an English Nursery Catalogue.)

Many people love to eat them fresh; others prefer them sliced and drizzled with honey. Red-fleshed Pêches de Vigne make exquisite fruit tarts and stunning, ruby-coloured jams. Their juice, which is delicious, has a deep, rich, garnet hue, like Cabernet Sauvignon or Merlot wine.

The Garon Valley in France is well-known for its agriculture. The main crops include the Pêche de Vigne, especially the red-fleshed type.

NAMES OF RED-FLESHED PÊCHES DE VIGNE

Named Pêche de Vigne cultivars exist. These are varieties that people have selected due to their superior characteristics.

Some names[17] of red-fleshed French Pêches de Vigne are:

L'attendue: This translates as 'The Expected', or the Long-Awaited'. Ripens early, at the beginning of July in France.

Sanguine: Translates as 'Blood'.

Sanguine Astruc: 'Astruc Blood'.

Sanguine Castelnau de Montmiral: Castelnau-de-Montmiral is a commune in the Tarn department in southern France.

Sanguine Chanas: 'Chanas Blood'

Sanguine d'Aout: 'August Blood'.

Sanguine de Besseas: Besseas Blood. A 'demi-précoce' peach, meaning 'semi-early'.

Sanguine d'été: Summer Blood. Large fruit, hardy, ripens in late August in France.

Sanguine de Juillet: July Blood. Sweet, juicy, ripe in mid-July in France.

---

17   *Information source: Pépinières Delay, Conservatoire d'Aquitaine*

Sanguine de Manosque: Manosque Blood.
Sanguine de Mauves: Mauves Blood.
Sanguine de Savoie: Savoie Blood. Fruit skin is dark red to burgundy, fuzzy. Flesh is beetroot-red and very dense, aromatic.
Musqué long en bouche: Long Musk Finish (literally 'Musk long in the mouth). A hardy variety, ripening at the beginning of September in France.
Sanguine de Touraine: Touraine Blood. Large fruit, hardy, eat when ripe, ripens in mid-September.
Sanguine Dufour: Dufour Blood.
Sanguine Durieux: Durieux Blood.
Sanguine Ferlay: Ferlay Blood.
Sanguine Latil: Latil Blood.
Sanguine Marcilly: Marcilly Blood.
Sanguine Peyrins: Peyrins Blood.
Sanguine précoce: Early Blood. Moderately vigorous tree, very juicy fruit, ripens in early August.
Sanguine Saint-Laurent: Saint-Laurent Blood. Vinous red flesh, good quality, ripens after mid-August.
Sanguine tardive: Late Blood. A pavie peach, i.e. clingstone. Large fruit, slightly downy skin, dense, red flesh, slightly discoloured around the kernel; juicy, tangy, used for pies and preserves, ripens early October.
Sanguine tardive gieusse: Gieusse Late Blood.
Sanguine vineuse: Vinous Blood.
Pavie sanguine tardive: Late Blood Clingstone.

Other Pêche de Vigne Names

The names of some Pêches de Vigne with yellow or white flesh include Tardive Montet, Jaune, Jaune Privat, Lajanié, Moïse, Perseque and Red Baron.

The Pêche de Vigne Festival

The Pêche de Vigne is celebrated annually in the town of Soucieu-en-Jarrest, south-west of Lyon, located in the department of Rhône in the region Rhône-Alpes. The citizens of Soucieu-en-Jarrest call it the capital city of Pêches de Vigne. The annual Feast of the Pêche de Vigne is held here in September.

The festival celebrates the peach in all its forms — fresh, jams, sorbets, juices etc. There is a contest for the best Pêche de Vigne cake or tart, a 'peach kernel race', a costume parade, craft and food stalls, street theatre, a funfair, musical bands and dancing.

Red-Fleshed Peaches in New Zealand

The Blackboy Peach

In New Zealand, red-fleshed peaches are known as Blackboy peaches. They are not generally sold in supermarkets; rather, they can be found—in season— at local farmers' markets. In the cooler climes of New Zealand they ripen in autumn, between March and late April. Notably, they grow on the Banks Peninsula on the South Island, latitude 43.7500° south and Greytown near the southern end of the North Island, 41.0667° south.

Leonard Lowe on his blog 'Paradise Enough' writes: 'They are a hardy, disease-resistant, late-fruiting peach, with a velvet skin that holds within it a delicate, sweet, juicy flesh like that of white-fleshed peaches, but with a complex flavour and deep purple hue reminiscent of fine red wine. Indeed, some commenters have used words like "raspberry," "spice" or "cinnamon" to describe some of the subtle complexities in this peach's flavour – terms not out of place in a wine connoisseur's notebook!

'Also known as "Pêche de Vigne," they originate from the Lyonnais region of France, where for centuries they were planted in vineyards to provide early warning of pests and diseases. While they can be found in New Zealand, they are rare in other parts of the world, including Australia, despite their magnificent colour, flavour, and usefulness — they are delicious fresh, cooked or preserved.'

Blackboy peaches should ripen and soften on the tree to give them the best flavour for eating fresh. If they are picked too early they can lack flavour.

The Blackboy peach has the true deep, purplish-red skin and flesh. It is juicy and fragrant, with an excellent flavour if picked when perfectly ripe. New Zealanders use them in pies, tarts, fruit muffins, peach chutney, peach sauce, jam, peach ice-cream, peach cake, peach wine or schnapps, peach liqueur and brandied peaches. They also bottle them in syrup, or freeze them stewed. The fresh fruit keeps well for several weeks when stored in the refrigerator.

New Zealand Blackboy peaches are usually propagated from seed because they grow fairly true to their parent tree. They are healthy and vigorous trees. Unlike modern white-fleshed and yellow-fleshed peach cultivars, they do not get leaf curl and they are said to be almost disease free. They also appear immune to serious pests.

They grow rapidly and start fruiting early. Within three years of sowing, seedling Blackboy peach trees will begin to flower and fruit. When mature, they bear fruit prolifically. The fruit ripens all at once, over a relatively short period. Trees are vigorous, and tolerate short dry spells.

Origins of the Blackboy Peach

The name 'Blackboy Peach' is unique to New Zealand. Exactly how these red-fleshed peaches reached the shores of the 'Land of the Long White Cloud' in the first place is interesting. The most likely scenario is that a sackful of kernels arrived with the early French settlers on board a sailing ship, in August 1840.

By the early nineteenth century the British had already colonised New Zealand's North Island. There were no French colonies established in the Pacific, so the French decided to annex the South Island.

French whaling ships were regularly sailing between France and New Zealand. The commander of one of these whaling vessels, Captain Jean-François Langlois, '...felt that Akaroa, on the Banks Peninsula, would make an excellent French naval base.'

The township of Akaroa is located on sheltered harbour. It is overlooked and surrounded by the weathered flanks of a miocene volcano. Volcanic soils are renowned for their rich fertility.

Langlois '... began forming plans to take the South Island for France. He negotiated with, and obtained signatures from twelve Ngai-Tahu Maori chiefs from Port Cooper, whereby he bought of most of Banks Peninsula, on the east coast of New Zealand.'[18]

For the purpose of colonising the South Island of New Zealand for France, Langlois formed a company called La Compagnie Nanto-Bordelaise. In 1839 King Louis Philippe approved of the project. The French government subsidised Langlois' company and lent him a passenger ship to convey French settlers.

---

18   Source: New Zealand in history. 'The colonisation of New Zealand—French colonists in Akaroa, South Island' by Robbie Whitmore; "Le colloque d'Akaroa, 16 - 19 août 1990", The Heritage Press. Article by Peter Tremewan; "The French in Akaroa" T.L. Buick.

'The delicate problem now was to annex the South Island without provoking the British, well installed in the North Island. It was hoped that the Commissioner of the King of France, Captain Lavaud, would simply be able to take the South Island in the name of France, but it was decided a more diplomatic solution would be to simply buy up land from the Maori. Future French settlers would be installed over the South Island, which would then eventually be claimed for France.

'However, by the time Langlois had been able to gain official backing from France, the Maori population of Banks Peninsula had increased considerably. This was mainly the result of the Ngai Tahu captives returning home from the North Island.'[19]

'A convoy of French settlers left the port of Rochefort in March 1840 on board the ship "Comte de Paris" (named after King Louis Philippe's grandson). One month earlier, the warship "L'Aube" ("Dawn") had set sail, under the command of Captain Lavaud. Captain Lavaud had instructions to represent the French Government in New Zealand until the arrival of a Governor.'

How excited the French colonists must have felt. Before them lay a new and promising land, one quarter the size of France, occupied by only three or four thousand Maori inhabitants. It must have seemed ideal. On the voyage they would have carried with them everything they felt they needed to establish their new life—including grapevines and Pêches de Vigne.

However the British, having got wind of French intentions, acted swiftly.

'... just one month before the "Comte de Paris" left France, the British signed the Treaty of Waitangi with Maori Chieftains, on 6th February 1840. The South Island Maori chiefs signed the treaty a little later, on 30th May of the same year. After the signing of the treaty, a British warship sailed to Akaroa and planted the Union Jack.'

---

19   ibid.

In those days news travelled at the speed of sailing ships.

'The unknowing French duly arrived at Akaroa in August 1840 to discover they would be settling in a British colony.

'The French also discovered that the land bought by Langlois had been resold again, several times since, as was often Maori custom. Some British settlers laid claim to certain areas of land which had originally been bought by Langlois. Fortunately, due to much diplomacy, no major incident arose from this event. The French Government requested the British Government to protect the rights of French landowners in New Zealand, and this was agreed upon in 1841.

'The French colonists settled in Akaroa as planned, but instead of a large South Island French colony, just two small towns of around 60 French inhabitants were established. In spite of the small number of French colonists, quite a few New Zealanders are today descendants[20] from these original French settlers from Normandy and from the Charente. A number of Akaroa streets today still carry French names.'[21]

The Charente (latitude 45.8333 °N) is a department in southwestern France where the Pêche de Vigne is still grown to this day.

'The Poitou Charentes area is the central part of western France. The climate of the Charentes is one of the mildest in France, and the coastal area of this region is the sunniest part of France outside the Mediterranean coastal areas. In the southern part of the region, extensive vineyards provide the grapes that are used in the production of Cognac and the famous local apéritif wine *Pineau des Charentes*.

---

20   The 'Comte de Paris Descendants Group' now exists, and its goals are to 'Preserve the heritage of the French and German families who landed at Akaroa from the ship Comte de Paris in August 1840. As a group they aim to encourage the preservation and knowledge of the heritage of the Akaroa families and celebrate the National days of both France and Germany as well as participate in the celebration of the annual Akaroa French Fest.'

21   Ibid.

'The coastal area, including the towns of La Rochelle, Rochefort and Royan, is popular for seaside tourism and also with yachtsmen; however many kilometres of the coastline are given over to oyster beds, oysters being one of the big local specialities. The coastal islands of Oléron and Ile de Ré are famous for their beaches and maritime environment. [22]

In fact, one of the specialty dishes of the region is 'Gurnard tartar with raisins and vineyard peach'. (See the 'recipes' section in this book.)

Langlois himself was from Normandy, which is why he would have selected people from that region — possibly including many of his own kinsmen — to be among the first settlers.

Perhaps Langlois also chose people from the Charente region as colonists because he felt their homeland was similar to the Banks Peninsula, with its long coastline and its mild climate. And perhaps, if we were to look at an old bill of lading from the voyage of the *Comte de Paris* in 1840, we might have seen, written in the ledger, the words 'noyaux des Pêches de Vigne, à chair rouge'[23]...

---

22   *About-France.com*
23   'à chair rouge' means 'with red flesh'.

Red-Fleshed Peaches in Africa

The Cape of Good Hope is a rocky promontory at the southern end of Cape Peninsula, South Africa.

In the early 17th century the Dutch trading port of Batavia, in the East Indies (now Indonesia), grew large and important. Merchant vessels regularly made the long voyage from the Netherlands to Asia. The managers of the Dutch East India Company needed to build a supply station at a midway point on the route, and they chose the Cape of Good Hope.

In 1652, a Dutch surgeon named Jan van Riebeeck landed at the Cape, commissioned with the task of building both a fort and farming community on the Cape.

One of van Riebeeck's tasks include planting a vineyard. At the time it was thought that the consumption of wine was effective in avoiding scurvy among sailors on long sea voyages.

In 1659 the first South African wine made from French Muscadel grapes was successfully produced.

Production during those early days was low, and the wine produced in the Cape settlement was only exported to the trading port of Batavia. Eventually the Dutch East India Company permitted settlers to buy land and grow wine grapes for their own consumption. As the market for Cape wine expanded, the company brought in a winemaker from Alsace, along with winemaking equipment and a cooper to make oak wine barrels. A winery was built and the South African wine industry began.

At 48.5° north, Alsace is probably too far north for the winemaker from that region to have been familiar with Pêches de Vigne as a vineyard helper. However between 1688 and the 1690s the Cape colony received an inpouring of French Huguenots, driven from France by the fear of religious persecution.

The Huguenots brought with them their viticulture and winemaking experience from their homeland. Most of them hailed from the southern and central parts of France—the home of Pêches de Vigne.

Thus, it is possible that red-fleshed peaches first arrived in Africa with the Huguenots, in the 17th century.

Red-Fleshed Peaches In The USA

The Indian Blood Peach

There are two main types of Indian Blood peach; a clingstone and a freestone. The Indian Blood Cling peach and Indian Blood Free peach, as they are known in North America, have a dark reddish-grey skin. Some have flesh that is a solid, burgundy-purple all the way to the pit. Others have mottled red flesh with scarlet, tiger-like stripes, or flesh of a deep, purplish-red color giving way to streakier, lighter, red and even yellow flesh near the pit. The catalogues of some plant nurseries in the USA advertise 'Indian Free' peaches that have white flesh with red streaks.

The differences can be explained by the fact that these trees are propagated from seed, which introduces the chance of tiny, spontaneous mutations with every successive generation. Add to that the variations that occur in fruit depending on the local terroir, and you will understand why there are so many different-looking Indian Blood peaches. Variables such as soil, nutrients, fertilizer, etc. change the colour of the peach.

There also appears to be a rare sub-type called the 'Black Indian Blood' peach, which is smaller than the standard Indian Blood, and was originally grown for making 'peach pickles'.

All of them, however, ripen in mid-September in the northern hemisphere.

Uses

Indian Blood peaches are good to eat fresh, 'out of hand', but in eighteenth century U.S.A. they were usually used for pickling and preserving.

History: The Europeans

'Historians believe that peach trees were first introduced into the colonial settlements of the United States by the French explorers in 1562 at territories along the Gulf coastal region near Mobile, Alabama, then by the Spaniards who established Saint Augustine, Florida in 1565 on the Atlantic seaboard.[24]

'The peach trees were planted from peach seed imported from Europe in an effort to establish a self-sustaining, agricultural fruit tree product to feed the colonists.

How did red-fleshed peaches reach the Americas from southern Europe?

Some people believe that Spanish settlers brought peaches to St. Augustine, Florida in 1565; others claim that they came from France to a Gulf of Mexico settlement in 1562 or to Louisiana in the 17th century.[25]

Since there is no history of red-fleshed peaches in Spain, it is more likely that they came from France.

Either way, peaches flourished in the New World and quickly spread.

History: The Native Americans

'American Indians spread the planting of the peach trees throughout vast territories by transporting the peach seed to other tribes that lived in the interior regions.

---

24   *'History of Peach Trees' by Arbor Solution Tree Service of Harvard MA.*

25   *It has even been postulated that peach seeds could have been brought to the Americas by the ancestors of the First Americans when they crossed from Siberia on the land bridge, during the last Ice Age. However, there is no archaeological evidence to support this theory.*

'This new crop of fruit was fast growing, producing a delicious peach two or three years from planting. The trees were so productive and vigorous that sometimes, widespread impenetrable thickets became established from the peach seeds that fell to the ground from fruit unharvested. The illusion was formed by settlers after 1600 that the peach trees were native to the United States, since they were so widely spread and grew so vigorously everywhere.

'Although peaches were commonly grown in Old World gardens and orchards, spreading from Asia on the tides of civilization, most European explorers mistakenly thought the peach was an indigenous American product. From Pennsylvania south, peach trees merged into the surrounding vegetation so completely that the earliest natural historians, even John Bartram, America's first great botanist, assumed the peach was a native tree.'[26]

Red-fleshed peaches became a favourite fruit of the Native Americans, particularly the Cherokee people, who planted peach pits extensively. They also let the pits fall to the ground, where they germinated and took root in the fertile soil and benevolent (to peaches) climate. These peaches grow true from seed, which is why the Native Americans were able to propagate them so successfully.

This resulted in the springing up of large tracts of peach groves, including red-fleshed peach trees. These groves were firmly established in the landscape by the time British and other European settlers arrived in the places that are now the southern states of the USA.

There were so many red-fleshed peach trees growing wild that settlers believed the trees were native to the USA and that they had always been cultivated by the Native Americans.

---

26 *Thomas Jefferson Encyclopedia. Reference Fruit: 'Peaches'. (C) The Thomas Jefferson Foundation. February 2003.*

The Success of Peach Trees

'Captain John Smith wrote about peach trees that were growing in Jamestown, Virginia in 1629. William Penn recorded in 1683 that dense, native thickets of wild peach trees were full of fruit just north of Philadelphia, Pennsylvania.

'The first plant nursery to become established in the United States was the Prince Nursery of Flushing, New York, in 1774 that sold grafted cultivars of peach trees to customers. General George Washington visited this nursery and had previously purchased fruit from them.

'By the beginning of the eighteenth century peach trees had naturalized so abundantly throughout the southeastern and mid-Atlantic colonies that John Lawson said they grew as luxuriantly as weeds: "we are forced to take a great deal of Care to weed them out, otherwise they make our Land a Wilderness of Peach-Trees."'[27]

'Aside from the trees cultivated by Native Americans and the naturalized Indian peaches, the European settlers in Virginia planted the fruit in enormous quantities, ultimately defining a distinctive American form of fruit growing quite different from the European tradition. Peaches were noted by John Smith in Jamestown as early as 1629, and in 1676 Thomas Glover said, "Here [in Virginia] are likewise great Peach-Orchards, which bear such an infinite quantity of Peaches, that at some Plantations they beat down to the Hoggs[28] fourty bushels in a year."

'Robert Beverley also described the "luxury of the peach" in early Virginian orchards: " ... some good Husbands plant great Orchards of [peaches], purposely for their Hogs; and others make a Drink of them, which they call Mobby, and either drink it as Cider, or Distill it off for Brandy."

William Bartram, the famous American botanist and explorer, wrote in his book, 'Travels', in 1773 several accounts

---

27   *Thomas Jefferson Encyclopedia. Reference Fruit: 'Peaches'. (C) The Thomas Jefferson Foundation. February 2003.*
28   *'... they beat down to the Hoggs..' i.ie. farmers would beat the branches of the fruit trees with sticks, making the fruit fall to the ground so that the pigs could eat it.*

of his observations of ancient peach and plum orchards growing in Georgia, South Carolina, and Alabama. Bartram visited the ruins of a French plantation in 1776 near Mobile, Alabama, and recorded, "I came presently to old fields, where I observed ruins of ancient habitations, there being abundance of peach and fig trees loaded with fruit."

Although Jefferson recorded the production of mobby in 1782 and 1785, it is difficult to determine whether he fermented it further into brandy.'[29]

THE FRENCH CONNECTION

'An extensive group of grafted peach trees was sent to the Thomas Jefferson fruit tree orchards from Prince Nursery.'

And here is where the French connection, begun in 1565, is strengthened.

'President Thomas Jefferson was instrumental in the importation of many new agricultural products from Europe through his influence as Minister to France before the American Revolution. The aggressiveness and monumental fruit production of peach trees impressed him to establish a living fence, that encircled his expansive gardens at his home at Monticello, Virginia, in 1794. Jefferson found many other uses for peach trees such as the brewing of brandy in 1782.

'Jefferson wrote to his granddaughter, Martha, in 1818 that "a slave is busy drying peaches for you". These sun-dried peaches were called peach chips and retained a good quality for eating, even after months of storage. Peaches were also juiced and mixed with tea to make a delicious drink.

---

29   *Thomas Jefferson Encyclopedia. Reference Fruit: 'Peaches'. (C) The Thomas Jefferson Foundation. February 2003.*

The 'Black Plumb Peach of Georgia'

In 1807 Thomas Jefferson planted in Virginia forty-one stones of 'The Black Plumb Peach of Georgia' (Indian Blood Cling Peach).

This naturalized peach had been planted throughout the State of Georgia by the Indians and was a dark-red velvety color with tiger-like striping. It was a fragrant peach, highly desirable because of its rich coloring and taste. It was also the ideal size to preserve by pickling, or by making into jams, preserves, cobblers, pies, cakes, and ice cream.

Jefferson believed that 'The Black Plumb Peach of Georgia' was a cross between naturalized peach trees and a French pêche de vigne cultivar called 'Sanguinole'.

'Black' Fruit

The word 'black' is used to describe Jefferson's red-fleshed Georgian peach as well as the New Zealand Blackboy peach. In the world of fruit, the word 'black' is often used to describe a colour that is dark purplish-crimson. Certainly, red-fleshed peaches are generally of a very deep, rich, purplish hue.

Wine grapes are commonly classified as 'black' or 'white'; viz. pinot noir and pinot blanc. 'Black' grapes are wine grapes that have a reddish or blue pigmentation in their skins. Pinot noir grapes are actually purple, and the wine they make is dark red.

Red wines typically fall into two different categories known as 'red fruit' and 'black fruit' flavors.

Black fruit is a general term for wine aromas and flavours that suggest blackberries, blueberries, black cherries, blackcurrants, or other 'black' fruits. In reality, no fruit is black. So-called black fruits are generally dark purple.

It is the same with flowers. What we know as black roses are actually really dark red roses. There are no truly black flowers; most that appear black are really very deep shades of purple or red.

Red-fleshed Peaches in 21st Century USA

In the USA, as in New Zealand, red-fleshed peaches are grown in backyards and small orchards. They are usually not commercially grown, or available in supermarkets. Some people say that this is due to the fact that their 'keeping qualities' are poor, and that once they are picked they should be preserved or consumed as soon as possible; either fresh, out-of-hand, or cooked. Keeping qualities, however, vary significantly among red-fleshed peaches and some will keep for several weeks if properly stored in the refrigerator.

Jenny Everett writes, 'Come summer in the South [of the USA], finding a fresh peach isn't much of a chore. There are a bushelful of varieties grown in Georgia and South Carolina alone. But one in particular is coveted by chefs and foodies: the red-fleshed Indian Blood Peach.

'"This is the peach all the in-the-know farmers' market junkies will be looking for in June and July," says Chris Hastings, chef ... in Birmingham, Alabama.

'What separates this peach from its fuzzy brethren is that when fully ripe, the Blood has a firm texture and is sweet yet slightly tart. Hastings considers it the single best peach for canning, pickling, and making chutney (though sinking your teeth right into the red-marbled flesh isn't out of the question either). But first, of course, you have to locate some.

'"The key is to ask the oldest person selling peaches you can find, and if they don't have them, they almost always know someone who has a tree on a corner of their yard and will bring a basket to you the following week," Hastings says. "It's very inside baseball, but well worth the effort."'[30]

---

30   'What's In Season: A Real Peach'. by Jenny Everett - June/July 2010

Freestone and Clingstone

As mentioned earlier, European settlers believed that red-fleshed peach trees had always been cultivated by the Native Americans. Thus they called them 'Indian Blood Peach'. Freestone varieties are known as 'Indian Free' or 'Indian Blood Free' peaches, and there is a clingstone type called the 'Indian Blood Cling' peach, which is said to be more tart in taste than the sweeter freestone type. The freestone type is also said to be resistant to leaf curl, unlike the Indian Blood Cling. Both types are said to tolerate clay soils very well.

Indian Blood Cling

Christine Dann in her April 12, 2014 article 'A peach by any other name…' says that the Indian blood cling peach is still grown in Monticello, the property in Virginia that used to belong to Thomas Jefferson. Monticello boasts a wonderful collection of heirloom plants, including red-fleshed peach trees descended from the peach tree Jefferson purchased from a Washington nurseryman in 1807.

Whether this peach is closely related to New Zealand's freestone Blackboy peach, or even to American freestone blood peaches, is unknown. Anecdotal evidence, however, indicates that the Indian Blood peach is very different from the Blackboy.

Indian Blood Free

'Indian [Free] peaches … are a very productive late season heirloom variety of freestone peaches. There are only two other known varieties of peaches similar to the Indian peach, the Sanguine de Manosque and the Sanguinole peach, both heirloom varieties from France. They share the red-flesh trait, yet each variety ripens at different times and varies in size and climate temperance.

'Indian [free] peaches are distinguished by dark red flesh, their fairly large size and their velvety thin skin with mottled layers of red tones throughout. Their tender firm flesh is marbled with ruby and crimson red streaks from the skin to the fruit's stone. Indian peaches' high acid content contributes to a tart, brisk mouthful with undertones of blackberry, plum and baking spices.

Freestone peaches are usually considered to be fresh-eating peaches, but Indian peaches are quite versatile and are well-known as excellent peaches for canning and baking.

They can be used in the following dishes:
- fresh fruit salads
- savory salads
- appetizers,
- sauces, coulis and purees
- preserves

Tom Conway of Tall Clover Farm in the USA writes: 'Indian Free peach flesh color can change depending on growing conditions each year; sometimes the flesh is white and red, and other times as richly red as a raspberry. One year the peaches were almost solid red.

'Here's a peach that pushes the envelope on being a peach. I first discovered it in my research to find another peach leaf curl resistant variety for my Pacific Northwest orchard. Because I don't spray my orchard and use organic practices, I look for fruit trees that can stand up to our incessant spring rains and cool summer climate.

'Reasons Why Indian Free Peach is one of my favorites!
- Thrives in cool clime of of coastal Pacific Northwest.
- Scoffs at peach leaf curl.
- Sports a thick, fuzzy mahogany brown skin.
- Enjoys freestone status and very firm flesh.
- Tastes sweet then tart, with strong overtones of blackberry and black cherry.
- Surprises first-time eaters with its rich red flesh and juice.
- Keeps well (so far for two weeks in fridge).

- Harvests very late (beginning in early October), but still ripens nicely.'[31]

The Indian Blood peach differs from the Pêche de Vigne in being robust and relatively disease resistant. By contrast the Pêche de Vigne is quite disease-susceptible. As mentioned, it was planted in vineyards to act as a warning of impending disease, so that vignerons could apply treatment before the grapes were badly affected.

---

31 'Indian Free Peach Delicious, Unique on All Levels' Tom Conway, Tall Clover Farm. Nov 11, 2008

Red-Fleshed Peaches in Australia

It is almost certain that red-fleshed peaches in Australia are descended from the French Pêches de Vigne.

China is the true home of peaches, and Australia's gold rush of the 1850s attracted many thousands of Chinese men—and a few women—to Australia.

Most of the men were indentured or contract labourers, who were making the long trip not to find a place to live, but to discover gold. They were not seeking to amass wealth for themselves personally, but instead to obtain money to send back to China, where their wives and families were waiting for them.

Soon after reaching the gold fields, many Chinese started other enterprises, including market gardening. Some of them may have carried plant material with them on the voyage to Australia, but it would hardly have been peach kernels. Anyone planning to grow food for survival in a harsh, new environment would plant herbs and vegetables which quickly matured and could be harvested within weeks, rather than trees that might take five to seven years to bear fruit. In any case, had anyone brought peach kernels, they would most likely have been from white fleshed peaches. Many sources confirm that yellow-fleshed peaches are more popular among Americans and Europeans, while Asians prefer the white-fleshed varieties.

Thus it is extremely unlikely that red-fleshed peaches were brought to Australia in the 19th century by the Chinese. However it is not impossible that by the time of the first gold rush, red-fleshed peaches had already arrived!

How long red-fleshed peaches have existed in Australia and exactly where they were first grown is a matter for conjecture; nonetheless some clues do exist.

Back in the 18th century Australia missed out on being colonised by the French by a matter of weeks. It was only when the British got wind of France's plans to claim the great southern land that they acted to claim it themselves.

The French in Australia

Museum Victoria's article on 'History of Immigration from France' reports, 'Maritime "superpowers" France and Britain were fierce colonial rivals in the South Pacific region during the great Age of Exploration. Motivated by scientific interest and trade, French explorers began arriving on Australian shores.

'The British, deeply suspicious about French intentions in the region, moved quickly to establish colonies in many parts of Australia.'

'Had it not been for a few misadventures, Australia might have been a French colony. 'The French—despite having claimed the western side of the continent, drawn detailed charts and made thousands of important scientific discoveries—did not stick around to develop the nation.'[32]

The First Fleet is the term denoting the eleven ships which left Great Britain on 13 May 1787 to found a penal colony that became the first European settlement in Australia.

'Following the arrival of the First Fleet in 1788, the first French settlers soon began to arrive, including officials, convicts and refugees. Over the ensuing decades, many French

---

32   *Australia could easily have been a French colony. November 12, 2012, By Aleisha Orr.*

settlers would go on to become land owners, merchants and wine-makers.'[33]

'The French were among the first non-British settlers in New South Wales. Many arrived as officials, convicts and refugees after the French Revolution. Many soon made a mark as landowners, businessmen, merchants and wine growers.'[34]

'More French immigrants arrived from the 1820s onwards. By 1871, there were almost 2500 France-born settlers in Australia, many attracted by the gold rush. Within 20 years, the community had increased to 4500 following the arrival of tradesmen, farmers, winegrowers and even escaped convicts from a French penal settlement in New Caledonia.'[35]

The French winemakers, especially if they hailed from the south of France, would have been likely to bring with them the seeds of the useful red-fleshed 'vineyard' peaches.

But by that time red-fleshed peaches might already have been growing in Australia. There is one state, Tasmania, which is the most likely place for the first red-fleshed Pêche de Vigne to have been planted by the French.

THE TASMANIA-FRANCE CONNECTION

Since the earliest years of European world voyages of exploration, Tasmania has had many French connections. This island state, the southernmost state of Australia, is strewn with French place names that date from before most of Tasmania's towns were even settled. The French were among the very first European explorers to visit Tasmania. They came, they discovered and they named, drawing maps of the great southern land and labelling them.

---

33   'History of immigration from France'. Museum Victoria. Retrieved 10th June 2015.
34   Ibid
35   'The France-born Community'. Australian Government Department of Social Services article on 'Settlement and Multicultural Affairs', 2nd April 2014

When British explorers navigated Australian waters and made inland forays, their purpose was to colonise the land. By contrast, the French were interested in intellectual pursuits — learning about the geography, the natural history and the people.

Because the French were more interested in scientific research than the British, they usually carried more scientists on board their vessels. Before the invention of photography they needed artists, too, to capture images of the landscapes and flora and fauna. The French expeditions that were not establishing settlements generally possessed better knowledge of agriculture and horticulture than the colonists. Scientists and artists aboard the French naval ships *Le Géographe* and *Le Naturaliste* included zoologists, botanists, geographers, mineralogists, a hydrographer, astronomers, painters, draughtsmen and — most importantly for us — gardeners.

An Eighteenth Century French Garden in Tasmania

Intriguingly, the French left more than just a legacy of names in Australia. They also left at least one garden.

The explorers would build gardens wherever they touched dry land. Danielle Clode, author of *Voyages to the South Seas: In Search of Terre Australes,* says, 'It was a common maritime tradition to leave plants and animals not only for other sailors but also for the native people to exploit... sort of a gesture of goodwill to subsequent sailors and the native people.'

In Tasmania's south, at Recherche Bay, 'On 4 February 2003, environmental activists located rows of moss-covered stones in dense bushland on the north-eastern peninsula of Recherche Bay in Tasmania. The stones formed a rectangle roughly 9 metres x 7.7 metres in size; this was further divided into four smaller rectangles and enclosed a "plinth" measuring 1.8 metres x 1.7 metres.'

This was almost certainly the site of a garden established by French expedition member Felix Delahaye in 1792.

'But who was Felix Delahaye and what do we know of his garden and his role in d'Entrecasteaux's expedition? Born in 1767, Felix Delahaye had worked with the Ecole Botanique of the Jardin du Roi in Paris under Andre Thouin (after whom Thoin Bay in Tasmania is named).

In 1791, France's National Assembly agreed to dispatch to New Holland (Australia) an expedition captained by Antoine-Raymond-Joseph Bruny d'Entrecasteaux.

'D'Entrecasteaux attracted many capable officers to his project. The party included two hydrographers, four naturalists, a mineralogist, two artists, two astronomers and our gardener Felix Delahaye.' [36]

Felix Delahaye was nominated for the expedition by Andre Thouin of the Jardin du Roi (the King's Garden at Versailles, mentioned earlier). After outlining his previous work experience in Rouen and Paris, Thouin went on to describe his young assistant to d'Entrecasteaux as

> "... strong, vigorous and well-suited for voyages. Gentle, honest and of an exact probity. Active, hardworking and passionately loving his calling. Knowing by theory and by practice the processes of gardening and knowing very well the plants cultivated in the Jardin du Roi."[37]

D'Entrecasteaux accepted Thouin's glowing recommendation and it appears that Delahaye left Paris for the port of Brest in 1791 in a stagecoach laden with seven cases:

> "4 of vegetable seeds, 1 of nuts of fruit trees, 1 of gardening utensils and the last of the gardener's clothes".[38]

---

36   *A French Garden In Tasmania. The legacy of Felix Delahaye (1767-1829). Edward Duyker.ctoria University Press, 2005.*

37   *Thouin quoted by L. Letouzay, ed., in Le Jardin des Plantes a la croisie des chemins avec Andre Thouin 1747-1824, Paris, Editions du Museum, 1989, p. 228.*

38   *Ibid., p. 229.*

Note that this list has been translated from the French. In the English translation the word 'nuts' has been used for the French 'noyaux'. A more apt translation might have been 'stone fruit pits'. In other words, the young gardener Delahaye departed on the voyage with a sack of stone fruit pits in his care—including, we may surmise, either the stones of Pêches de Vigne or peach pits from the King's Garden where 'Peches Violettes' (Purple Peaches) grew, or had once grown.

Along the way the expedition called in at Cape Town, where French migrants already resided, including winemakers from regions familiar with the Pêche de Vigne.

The expedition reached Van Diemen's Land (Tasmania) on 21 April 1792 and two days later discovered Recherche Bay, where fresh water was found.

It was during this visit to Recherche Bay that Delahaye planted his garden. According to d'Entrecasteaux: 'Various seeds sowed by M. La Haye, gardener-botanist, might in future furnish supplies to navigators who will shelter in this haven ...'[39]

Delahaye's principal duty during the expedition was to collect useful seeds and seedlings, but in his instructions one can gain some understanding of why he planted a garden at Recherche Bay:

> 'If the gardener is provided with fruit trees and plants of an economic nature either in Brest, the Canaries or at the Cape of Good Hope, he will not fail to plant some of these vegetables in the places which appear to him to be most favourable to their multiplication in New Holland and choose localities where it is probable that European vessels will be able to call. But above all, he is urged to deposit them in the hands of some inhabitants of Botany Bay, choosing people who will attach value and who will take care to propagate them. This will

---

39   D'Entrecasteaux, *Voyage to Australia and the Pacific*, p. 38.

always be to advance towards the object proposed: to deposit at the entry to the South Seas, our useful vegetables which, multiplying there (with some hope of success because of the similar climate), will one day be transported to the islands and on to the continents of this great part of the world. It will be good to leave in this place the seeds of all our species of vegetables that the gardener takes with him, as well as the nuts and seedlings of our fruit trees.' [40]

'The expedition never called at Botany Bay and it was by no means certain that it would return to Van Diemen's Land. The garden, therefore, was probably planted with altruistic motives rather than as an investment in future fresh provisions for the expedition itself.

The D'Entrecasteaux expedition, in particular, was responsible for many Tasmanian place-names because it remained in Tasmania for a long time, doing extensive charting and mapping and naming of bays, harbours, peninsulas and other formations along the coastline.

'In December, once more D'Entrecasteaux sailed to Recherche Bay on the south-eastern coast of Van Diemen's Land to replenish. In his account d'Entrecasteaux tells us of their return visit to the garden site: 'The garden had not been a success; nothing, or nearly nothing, had grown. Was it because the season was not favourable, or because the seeds that had been sown had deteriorated? I instructed M. La Haye, gardener-botanist, to proceed to the spot and try to discover the cause.[41]

After yet another visit to the garden—this time in the company of one of the indigenous inhabitants—d'Entrecasteaux provided more detail: "... a few chicory plants, cabbages,

---

40 *"Instructions pour le Jardinier de L'Expedition autour du monde de M. d'Entrecasteaux"*, pp. 232-233.
41 D'Entrecasteaux, *Voyage to Australia and the Pacific*, p. 140.

sorrel, radishes, cress and a few potatoes had grown, but had only produced the first two seminal leaves".[42]

In his journal Delahaye blamed the fate of the garden on the "drought" and remarked that the earth was very hard.[43] To d'Entrecasteaux, however, he apparently declared that the lack of success of the garden was owing "to the seeds having been sown in too advanced a season".[44]

The fact remains, however, that it was not an absolute failure; Delahaye tells us that he picked a number of very small potatoes and attempted to explain to the Aboriginal people "that this root cooked simply on embers was good to eat".[45]

The peach pits sown in Tasmania by the French may not have survived, but there were many more seeds carried by European sailors and planted on Australian shores.

Captain Cook visited Adventure Bay on Bruny Island in 1777 on board the HMS Resolution. William Bligh was his sailing master. Bligh returned to Adventure Bay with the botanist Nelson in 1788. As was the practice of many mariners, these men planted a number of small fruit trees and vines as a future food source. The plants had been brought from the Cape of Good Hope.

Whether or not red-fleshed peaches originally arrived in Australia directly from France, brought by early European explorers or settlers, or whether they reached this country via a more circuitous route from New Zealand or Cape Town, we may never know.

In the nineteenth and early twentieth centuries, as now, there was much commerce between Australia and New Zealand. In those days strict customs and quarantine laws did not yet exist, and plant material was swapped freely between countries.

---

42   *Ibid., pp. 141-142.*
43   Delahaye, *"Journal"*.
44   D'Entrecasteaux, *Voyage to Australia and the Pacific, p. 142.*
45   Delahaye, *"Journal"*.

However in addition to the French clues, anecdotal evidence points to the idea that Tasmania, for whatever reason, could well have been their first point of entry. In Tasmania red-fleshed peaches are called Blackboy Peaches, a clue which rather indicates that the trees which flourish and grow there in the 21st century may have found their way to Australia in the pockets of people travelling from New Zealand.

For example, the red-fleshed peach trees at our farm on the Mornington Peninsula in Victoria are descended from pits given to them by a Tasmanian gardener.

'These are from a Blackboy Peach tree,' he stated. 'They came from an orchard up in Tasmania's north-west, in the Ulverstone-Spreyton region. They have to be grown from pips. You don't graft them.'

When in 2011 Tasmanian Leonard Lowe posted an article about Blackboy Peaches on his blog 'Paradise Enough', he attracted numerous comments from people in mainland Australia. One person wrote that she wanted to grow a Blackboy peach tree in Victoria but was having trouble finding one. She grew up in New Zealand where her granddad nurtured a Blackboy peach tree in his orchard, and she would like her Australian husband and child to taste them.'[46]

Another person wrote: 'Am I right in understanding that you have a Pêche de Vigne or Black Boy [sic] Peach here in Australia? I have been trying to find one for nearly seven years, since returning from living in Lyon France. ... This really is exciting.'[47]

This indicates that red-fleshed peaches were growing in Tasmania long before they reached mainland Australia.

---

46   *Posted by Janette | July 9, 2011*
47   *Posted by Tamz | October 23, 2011*

Rare and Heritage Fruit in Australia

Many of the rare and heritage fruits that exist in Australia today are clonally descended from plants brought by the early European settlers, when few, if any, quarantine laws existed. Good luck rather than good stock monitoring limited the number of plant diseases unintentionally imported during the early days of colonization. Fortunately, by 1879 it was recognised that in order to prevent the introduction of serious pests and diseases, quarantine measures were needed. In 1908, the Commonwealth Quarantine service came into operation and took over local quarantine stations in every Australian state.

However, before 1879, there was no limit to the varieties of fruiting plants that could be imported into Australia. Many of those old genetic lines survive to this day but sadly, many others have been lost.

Over the course of the decades since 1879 Australian fruit growers imported (through quarantine) the latest new cultivars bred by overseas agricultural research stations. Year by year, as scientific advances in breeding and genetics were made, the older cultivars fell out of fashion and were swept aside in favour of the new. They, too, became part of an almost forgotten fruit inheritance.

Red-fleshed Peaches on the Mornington Peninsula

As mentioned earlier, Australia's Mornington Peninsula in Victoria is ideally located for peach growing, at 37.8° south of the equator. Its maritime influence contributes to its unique "terroir".

Red-fleshed peach blossom on the Mornington Peninsula

Famous for its apple orchards, cherry orchards, strawberry farms and vineyards, Australia's Mornington Peninsula has many sites with the perfect conditions for stone fruit growing.

Within the low, rolling hills of the Peninsula there is considerable diversity of climate, soil types, rainfall and altitude as well as the usual variations in topography, which give each vineyard and orchard site its unique characteristics, or 'terroir'.

In particular the Red Hill Ward with its rich, red soils, grows excellent peaches. This fertile soil developed on tertiary basalts from ancient volcanic lava flows. Millions of years ago, this entire landscape was aflame with volcanos spewing mineral-rich material from beneath the earth's crust. Over the millennia, wind and rain eroded this material into the vibrant, red fruit-growing soils we know today.

One estate in particular specialises in growing red-fleshed peaches. Situated on the peninsula's eastern side in a well-watered valley with gentle, north-facing slopes, it is the farm cherished by the writer of this book.

In late summer the long-awaited red-fleshed peaches ripen at last, and are snapped up by friends, family and eager customers at local fruit and vegetable shops.

About Pleaches

New fruit cultivars continue to arise in the 21st century. 'Pleaches' are not true peaches, but a red-fleshed cross between a peach and a plum. The fruit is solid wine-red all the way through and the taste is tart but delicious, like raspberries.

From AAP, February 21, 2010, an article entitled 'Mudgee Farmer Bruce Davis Creates New Fruit':

'Is it a plum? Is it a peach? It's probably a pleach as it's a morph of the two tasty stone fruits. Whatever it is, it's a love child of the two, accidentally created by a retired NSW farmer.

'Bruce Davis from Mudgee in the state's central west couldn't believe it when he discovered he had grown a cross between a peach and a plum. The fruit looks like a peach from the outside, but resembles a red plum when bitten into. 'The unusual fruit is believed to be the first of its kind ever grown in the state.

'Mr Davis grows peach and blood plum trees alongside each other and believes the peach/plum tree may have grown from compost that contained plum seeds.

'"It's a really interesting piece of fruit and it's very tasty," Mr Davis said.

'A cross between a plum and an apricot, known as a pluot, has been grown in the past, but a peach and a plum is a new combination for NSW, Primary Industries Minister Steve Whan said.

'Industry and Investment NSW Mudgee horticulturist Susan Marte said this was the first time she had heard of anyone accidentally crossing the two fruits.'

Red-Fleshed Peach Recipes

In this section we present a collection of recipes for red-fleshed peaches. Any red-fleshed peach can be used—Pêche de Vigne, Blackboy, Indian Blood Free, Indian Blood Cling, Blood Peach, Nectavigne®, or even Pleach. White-fleshed or yellow-fleshed peaches can also be substituted.

Note that red-fleshed peaches combine well with:
- other stone fruits
- honey
- egg custards
- lavender
- lemon or orange
- cardamom
- basil, rocket (arugula)
- cayenne and chillies
- almonds, hazelnuts and pistachios
- vanilla
- white chocolate
- rosemary

How to Peel Peaches

Be aware that if your peaches are not fully ripe, this method of peeling them will not work. If they are still hard, leave them to soften for a couple of days before using them.

To peel peaches, dip them in a saucepan of boiling water for 20 to 45 seconds. Remove peaches from the boiling water using a slotted spoon, and plunge them straight into a large bowl of iced water for a few minutes.

Notch the skins with a sharp knife. They should slide off easily.

Red-fleshed Peach Desserts

RED-FLESHED PEACH CLAFOUTIS

A French recipe, '*Clafoutis à la pêche de vigne*'.
Vegetarian. Serves 4 people.

Ingredients
- 4 red-fleshed peaches
- 2 + 2 egg yolks
- 100 ml (3 oz) milk
- 200 ml (6½ oz) heavy whipping cream
- 100g (6½ tablespoons) hazelnut meal
- 50g (3½ tablespoons) brown sugar
- 20g (4 teaspoons) plain (all-purpose) flour
- 15g (1 tablespoon) salted butter

Instructions

Preheat oven to 180 °C (350 °F, gas mark 6).

Rinse and dry the peaches. Cut them open, remove the pits and slice the peaches thinly.

In a bowl, whisk the eggs and yolks with the brown sugar. Pour in the milk and the cream.

Stir in flour and 50g of the hazelnut meal.

Grease a pie or flan dish generously. Line the sides with the remaining hazelnut meal and arrange the peach slices on the bottom of the pie dish.

Pour the clafoutis batter over the fruit.

Bake for 30 minutes and serve warm or cold.

Tip: You can also use mini tart dishes for individual clafoutis.

RED-FLESHED PEACH COBBLER

Spectacular, delicious and really easy to make. Vegetarian. Serves 6 people.

Ingredients
- 6 or 8 red-fleshed peaches
- 50g (3½ tablespoons) butter
- ¾ cup plain (all-purpose) flour
- 2 teaspoons baking powder
- pinch of salt
- ¾ cup milk
- 80g (5½ tablespoons) castor sugar

Instructions

Preheat oven to 180 °C (350 °F, gas mark 6).

Place butter in a casserole dish or pie or flan dish and melt for 20 seconds in microwave.

Put flour, baking powder and half the sugar into mixing bowl with pinch of salt. Combine, then add milk and beat to a smooth batter.

Pour batter into casserole dish on top of the melted butter. Do not mix.

Thinly slice fruit and arrange on top, as if the batter is studded with red jewels. Then sprinkle fruit with remaining sugar.

Bake till the edges have begun to caramelise and top is golden brown — it usually takes around 40 minutes.

Serve with cream or ice-cream.

NO-BAKE RED-FLESHED PEACH PIE
Vegetarian

Ingredients
For the crust:
- ¾ cup crumbs of sweet biscuits such as graham cracker (US), wheatmeal biscuits (Australia) or digestive biscuits (UK)
- ⅛ teaspoon salt
- 1 tablespoon unsalted butter
- 100g (3 ½ oz) white chocolate, finely chopped
- Cooking oil spray such as rice bran oil

For the filling:
- 140g (5 oz) low fat cream cheese, softened
- ⅓ cup icing sugar (confectioner's/powdered sugar)
- ½ teaspoon vanilla extract
- 2 cups frozen fat-free whipped topping, thawed. Or use sweetened whipped cream instead.

For the fruit topping:
- ⅛ teaspoon salt
- 2 tablespoons peach jam [see page 100]
- 1 tablespoon peach schnapps
- ½ teaspoon fresh lemon juice
- 3 medium sized peaches, peeled [see page 63] and pitted and cut into 1 cm (½") wedges

Instructions
To make the crust:

Coat the base and sides of a 23cm (9") pie plate with cooking spray.

Place crumbs and ⅛ teaspoon salt in a food processor and process until well-blended.

Put butter and chocolate in a microwave-safe bowl. Microwave at high for 30 seconds, then remove from microwave and stir until chocolate is smooth.

Add to melted chocolate to food processor and pulse until combined with crumbs.

Press mixture into the base and sides of the oiled pie plate. Place in the freezer and freeze 20 minutes or until set.

To make the filling:
Combine cream cheese, sugar and vanilla in a medium sized bowl and beat with an electric mixer at medium speed until smooth.

Gently fold in whipped topping or sweetened whipped cream.

Carefully spread filling over bottom of crust.

To make the fruit topping:
Place peach jam in a large microwave-safe bowl; microwave in high for 30 seconds or until it starts to bubble.

Stir in remaining $\frac{1}{8}$ teaspoon salt, schnapps, and juice.

Add peach wedges and toss to combine.

Arrange the peach wedges decoratively over pie.

Chill for 3 hours before serving.

RED-FLESHED PEACH PARFAIT WITH ROSEMARY AND CREME ANGLAIS

A French recipe.

Ingredients
10 - 12 red-fleshed peaches
200g sugar
250 ml water
1 sprig fresh rosemary

For the topping:
2 gelatine leaves
reserved juice from the cooked and strained peaches

For the Creme Anglaise:
70g sugar
6 egg yolks, beaten
500 ml of milk 1
11g vanilla castor sugar
1 heaped tablespoon cornflour for thickening

Special equipment:
Parfait glasses
Microwave oven

Place the cornflour in a bowl, then gradually add cold milk a little at a time, stirring constantly until cornflour is blended.

Add the rest of the Creme Anglais ingredients and mix well.

Cook over a medium heat for 8-10 minutes, stirring constantly, until it thickens.

Pour the cream into the parfait glasses. Set glasses aside in a cool place and allow the cream to cool down.

Meanwhile, peel your peaches [see page 63], cut into pieces and put them in a saucepan with the sugar, water and rosemary.

Bring peaches to the boil, then reduce the heat to minimum and let them reduce to a syrup. Stir frequently. It should take 15-20 minutes. Peaches are cooked when they are soft but do not stick to the pan.

Allow to cool, strain the juice through a strainer, remove the rosemary sprig.

Place the gelatin in a bowl with a little cold water, and alow it to soften.

Bring out the glasses of cream and place a good layer of cooked peaches on top.

Put the peach juice and both softened gelatin sheets into a microwaveable jug. Heat for 30 seconds in the microwave.

Then pour the syrup over the dessert, making sure you cover the entire surface of peaches. Refrigerate.

To serve, decorate with your choice of peach leaves, biscotti and whipped cream.

RED-FLESHED PEACHES IN ROSEMARY SYRUP

A French recipe. Serves 4 people.
Vegetarian, gluten-free.

Ingredients
- 6 red-fleshed peaches,
- 2 litres (8 ½ cups) water
- 350 grams (12 oz) of crystal sugar,
- 50g (2 oz) of rosemary honey (or other honey)
- 4 sprigs fresh rosemary.

Special equipment
- a bowl of cold water with plenty of ice-cubes in it
- a large jar with a tight-fitting lid

Instructions

Boil water with sugar and honey for 5 minutes.

Reduce heat, add 3 sprigs of rosemary and cook over medium heat for 15 to 20 minutes.

Meanwhile, boil another pot of water.

Make a small, cross-shaped incision on the skin of each peach and immerse them for 30 seconds in the boiling water. Then plunge them into the bowl of ice-water.

Peel off the loosened skin and place them in a jar.

Add 1 sprig fresh rosemary. Cover the peaches with syrup, close the jar tightly and let it cool.

The peaches will keep for several weeks. They can also be eaten immediately, while lukewarm.

GRILLED RED-FLESHED PEACHES WITH CHOCOLATE FILLING

Vegetarian.

Ingredients
- ½ cup crushed amaretti cookies
- 2 tablespoons brown sugar
- 4 large, ripe red-fleshed peaches, halved and pitted
- Cooking spray
- 8 teaspoons butter
- 30 g (1oz) bittersweet cooking chocolate, shaved

Instructions
Preheat griller (broiler).

Combine cookie crumbs and sugar in a small bowl.

Hollow out the middles of the peach halves using a melon baller.

Fill each peach half with 1 rounded tablespoon cookie crumb mixture.

Arrange peaches in an 30 x 18 cm (11 x 7-inch) glass or ceramic baking dish coated with cooking spray.

Place 1 teaspoon butter on top of each filled half.

Grill/broil for 2 minutes or until butter melts.

Sprinkle evenly with chocolate.

Allow to cool for a few minutes before serving.

RED-FLESHED PEACH BAKED CUSTARD

Vegetarian, gluten-free

Ingredients
- 6 red-fleshed peaches
- 150 ml cream with 18% fat content (eg. single cream, table cream)
- 5 egg yolks
- 150g (5 oz) sugar
- butter to grease the dish

Instructions

Pre-heat the oven to 210 °C (410 °F).

Wash and peel the peaches [seepage 63]. Slice them in half and remove the stones. Use absorbent paper to gently dry the fruit.

Put the fruit into a well-buttered baking dish.

In a large mixing bowl beat together the yolks and sugar.

Add the cream and keep beating.

Pour the egg mixture over the fruit and bake in the oven for 20 minutes.

Serve cold or slightly warm.

RED-FLESHED PEACH COMPOTE

Vegetarian, gluten-free

Ingredients
- 6 red-fleshed peaches
- 1 vanilla pod
- 1 sprig of thyme
- 1 tablespoon of olive oil
- 30g (1oz) butter
- 150g (5oz) brown sugar
- 6 tablespoons of sweet Muscat-based wine

Instructions

Peel the peaches [seepage 63], remove the stones and slice them thickly.

Slice the vanilla pod lengthways and scrape out the tiny black seeds. Mix them with the olive oil.

Heat the oil and butter on a low heat in a sauté pan (a frying pan/skillet with straight sides). Add the peaches and fry them quickly on all sides.

Add the vanilla pod, the sprig of thyme, the brown sugar and the Muscat wine. Combine well.

Cook on a low heat until the fruit softens, stirring now and then to make sure the compote does not stick.

When it is ready, take out the vanilla pod and sprig of thyme and set the compote aside to cool.

Serve warm or cold.

Store any uneaten portion in the refrigerator.

Red-fleshed Peach Iced Desserts

RED WINE RED-FLESHED PEACH SORBET
Vegan, gluten-free.

Ingredients
- 1 cup sugar
- 1 cup water
- 2 red-fleshed peaches, sliced
- 1 ½ cups Merlot red wine

Equipment
- an ice-cream maker or food processor

Instructions

Bring water and sugar to a boil in a saucepan.

Add peaches and stir well until sugar is completely dissolved.

Reduce heat to a simmer.

When the peaches have softened, crush them with a fork or spoon to release their flavour.

Cover the saucepan lightly with a lid and simmer for 25-30 minutes.

Remove from heat and allow to cool.

Pour the mixture through a strainer into a measuring jar, to separate the peach pulp from the syrup. The sweet peach pulp can be eaten as a dessert on its own!

Measure the quantity of syrup you have made. It should be around 1 ½ cups.

Mix the syrup with an equal quantity of Merlot red wine.

Process the mixture in an ice-cream maker.

If you do not have an ice-cream maker, freeze it then blend the frozen mixture in a food processor before re-freezing it.

Serve scoops of dark pink sorbet in small bowls, garnished with fresh mint leaves.

EASY RED-FLESHED PEACH SORBET
Vegan, gluten-free.

Ingredients
- 6 red-fleshed peaches, fully ripe.
- 3/4 cup sugar
- 1 teaspoon lemon juice
- pinch of salt
- water

Equipment
- saucepan
- sharp knife and slotted spoon
- bowl of cold water with ice blocks floating in it
- freezer bags
- food processor

Instructions
Peel and slice peaches. [see page 63]

Freeze peaches until firm.

Stir sugar into ¾ cup cold water to make a syrup.

When all the sugar is dissolved, pour syrup and frozen peaches into food processor with lemon juice and salt.

Blend until creamy and smooth. Serve straight away or freeze for up to three days in a plastic freezer container.

EASY RED-FLESHED PEACH ICE-CREAM
Vegetarian, gluten-free

Ingredients
- 300ml chilled thickened cream
- 1-2 tablespoon of icing sugar
- 300-400g red-fleshed peaches

Instructions
Slice the peaches into chunks about the size of strawberries.

Place them on a tray and leave them in the freezer until they are frozen through.

Put the frozen chunks into a food processor with the cream and sugar.

Process until smooth and creamy.

Serve immediately.

RED-FLESHED PEACH GELATO
Vegetarian, gluten-free

Ingredients
- 4 cups peaches, peeled and sliced [see page 63]
- 2 ½ tablespoons water
- 4 egg yolks
- 2 cups milk
- ¾ cup sugar
- 1 cup heavy whipping cream
- 1 tablespoon peach schnapps or Champagne (optional)

Special Equipment
- cooking thermometer
- ice cubes
- ice-cream making machine

Instructions

Put peaches and water in a large frying pan and set over a medium heat.

Cook, uncovered, until tender, stirring occasionally, particularly towards the end. Cooking will take 10 to 12 minutes,

While peaches are cooking, beat egg yolks in a medium sized bowl, then set aside.

Pour cooked peaches into a food processor or blender. Process to make a puree, then set aside.

In a small saucepan, heat milk to 80° C (175° F). This takes 7 to 10 minutes. Stir sugar into warm milk until dissolved, then remove saucepan from stove. Turn the heat to low.

Fill a bowl with cold water and add plenty of ice-cubes to make ice-water. Set aside.

Whisk a cup of warm milk mixture into egg yolks to make a custard mixture.

Return custard to saucepan, whisking all the time.

Replace saucepan on the stove. Cook and stir custard over low heat until mixture is slightly thickened. This can take around 15 minutes.

Remove saucepan from heat and cool quickly by placing it in the bowl of ice-water. Allow to cool down to 2 minutes, still stirring constantly.

Fold in heavy cream, peach puree, and liqueur or champagne if you wish.

Press waxed paper onto surface of custard or cover with lid (this helps to stop a skin from forming).

Refrigerate for several hours or overnight.

Fill cylinder of ice cream maker two thirds full of the mixture and freeze it according to manufacturer's directions.

Refrigerate the remaining one third of the mixture until you are ready to process it in the ice cream maker.

Transfer gelato to your freezer and allow it to freeze for at least 2 hours before serving.

Red-fleshed Peach Flans and Tarts

RED-FLESHED PEACH FLAN
Vegan

Ingredients
- approx. 250g red-fleshed peaches, peeled [see page 63] and sliced
- 1 cup rolled oats
- ¼ cup almonds, skins on, slightly smashed
- 1 cup wholemeal flour
- ⅔ cup brown sugar
- 2 teaspoons baking powder
- ¼ teaspoon salt
- ¼ cup walnut pieces
- ¾ cup safflower or sunflower oil[48], plus a little extra.
- ¾ cup soy milk
- 1 tablespoon finely ground linseed beaten into 4 tablespoons water
- 1 heaped teaspoon ground cinnamon

Instructions

Grease a 21-22 cm flan dish with the little extra safflower or sunflower oil.

Heat the oven to 180 °C (350 °F).

Place the oats and smashed almonds into a food processor and process them until well-blended.

Add to this mixture the flour, sugar, baking powder and salt. Process until thoroughly combined.

Pour the mixture into a large bowl.

Add the walnut pieces and ¾ cup oil.

Mix well until the all ingredients are well-moistened.

---

48   ...or any other flavourless vegetable oil.

Pour half the mixture into the bottom of the oiled flan dish then spread it out evenly and press it down with the back of a spoon to make a firm base.

Arrange the peach slices evenly on top of this base.

Mix the soy milk, linseed and cinnamon into the other half of the flour-and-almonds mixture in the large bowl, and pour all of it over the peaches.

Put the flan dish into the oven. Bake for 30-40 minutes, until, when you pierce the top layer with a metal skewer, it comes out clean.

Just before serving, dust the top with icing sugar.

This flan can be served cold, or warmed up as a dessert.

RED-FLESHED PEACH FRANGIPANE TART
Vegetarian. Serves 8 people.

Ingredients
For pastry dough
- 1 cup plain (all-purpose) flour
- 3 tablespoons sugar
- ½ teaspoon salt
- 6 tablespoons chilled, unsalted butter cut into 1 cm (½") pieces
- ½ teaspoon finely grated fresh lemon zest
- 2 large egg yolks
- ½ teaspoon vanilla
- 1 ½ teaspoon water

For frangipane filling
- 200-220g (7 to 8 oz) almond paste (not marzipan or almond filling)
- ¼ cup unsalted butter, softened
- 3 tablespoons sugar
- ⅛ teaspoon almond extract
- 2 large eggs
- 3 tablespoons plain (all-purpose) flour
- ½ teaspoon salt
- 560 g (1 ¼ lb) firm, ripe, red-fleshed peaches

For glaze
- ⅓ cup peach jam [see page 100]
- 2 tablespoons water
- 1 tablespoon almond liqueur (optional)

Special equipment
- a pastry or dough scraper
- pastry weights or dry chickpeas or raw rice.
- a fluted flan or tart dish 2 ½ cm (1") deep, with a removable base. 30 x 20cm (11 x 8") rectangular or 20cm (11") round.
- Pastry brush

Instructions

Make pastry shell

Put oven rack in middle position and preheat oven to 190 °C (375 °F).

Place flour, sugar and salt into a food processor and pulse until well combined.

Add butter and lemon zest and pulse until mixture resembles coarse meal with some small (about the size of a pea) lumps of butter.

Add yolks, vanilla and water. Pulse just until dough comes together and starts to form large lumps.

Turn out dough onto a work surface and divide into 4 portions.

Push each portion once with heel of your hand with a forward movement to help disperse the butter-fat.

Gather dough together using scraper and form it into a ball, then flatten it into a rectangle or circle to match your flan dish.

Put dough in flan dish and pat out with well-floured fingers into an even layer over bottom and up sides so it extends about ½cm (¼") above rim.

Put into the refrigerator and chill for 30 minutes.

Lightly prick tart shell all over with a fork, then line with foil and fill with pie weights.

Bake shell until it turns golden around the edges—about 15 minutes.

Carefully remove foil and weights and bake until pastry shell is golden-brown all over—about 15 minutes more.

Do not turn off the oven. Take out the cooked pastry shell. Leave it in the flan dish and place it on a cake-rack to cool down.

Make frangipane filling:

Place almond paste, butter, sugar, and almond extract into a bowl and beat with an electric mixer at medium-high speed until creamy. This takes around 3 minutes.

Reduce speed to low and add eggs, one at a time, beating well after each addition.

Mix in flour and salt.

Assemble the tart:

Cut the peaches in half, remove the stones and slice them into ½cm (¼-inch) wide wedges.

Spread frangipane filling evenly inside the tart shell. Attractively arrange peach wedges in the filling, skin sides down, taking care not to push too deep into filling.

Put the flan dish into the preheated oven and bake tart until frangipane is puffed and golden and edges of peaches are golden brown. This takes around 1¼ hours.

Make glaze:

Heat peach jam and water in a 1 litre (1 quart) saucepan over moderately high heat, stirring constantly, until peach jam has liquefied.

Remove from heat and push through a fine-mesh sieve into a small bowl.

Discard solids (or use them as a topping for ice cream!)

Stir in almond liqueur (if you are using it).

With pastry brush, generously coat top of hot tart with glaze.

Allow to cool in pan on rack for 15 minutes. Remove side of pan and cool tart completely for about 2 hours.

RED-FLESHED PEACH CARAMEL TART
Vegetarian. Serves 8 people

Ingredients

- 200 g (7 oz or 1⅓ cups) plain flour
- 125 g (4 oz) unsalted chilled butter, finely diced
- 2 tablespoons icing sugar (confectioner's sugar)
- Finely grated zest of 2 oranges
- 1 egg yolk combined with 1 tablespoon iced water
- 1 teaspoon vanilla extract
- 6 red-fleshed peaches
- 165g (¾ cup) castor sugar

Pastry cream
- 500ml (2 cups) milk
- 1 teaspoon vanilla extract
- 50g (⅓ cup) plain flour
- 75g (⅓ cup) castor sugar
- 1 x 5g titanium-strength gelatine leaf[49]
- 4 egg yolks
- 30g chilled, unsalted butter, diced

Special equipment
An 11cm x 34cm (4" x 13") rectangular tart tin with removable base.

---

49   *Gelatine Equivalents*
· *1 teaspoon Gelatine Powder = 3.3gm*
· *1 Gold Leaf = 2.2gm*
· *1 Titanium Leaf = 5gm*

Instructions

To make the pastry shell:

Place flour, butter, sugar, half the zest and a pinch of salt into a food processor and process until mixture resembles breadcrumbs.

Add yolk mixture and vanilla. Process until mixture just starts to clump together.

Form into a rectangle, wrap in plastic cling film and refrigerate for 30 minutes.

Roll out pastry on a sheet of lightly floured baking paper until 4mm (3/16 -inch) thick.

Line the tart tin with pastry. Trim pastry 5mm (¼ -inch) above rim of tin.

Freeze tart shell for 1 hour.

Meanwhile, preheat oven to 180 °C (350 °F).

After tart shell has been in the freezer for 1 hour, line it with baking paper, fill with dried beans or rice to weigh it down, place on an oven tray and bake for 15 minutes.

Remove paper and beans, and bake for another 15 minutes or until golden and dry.

To make the pastry cream:

Gradually bring milk and vanilla to the boil in a small saucepan.

Combine flour and sugar in a bowl, then add milk and whisk until smooth.

Return mixture to saucepan and stir until thick, then simmer, stirring constantly, for a further 3 minutes.

Soak gelatine leaves in cold water for 5 minutes, then squeeze out excess water.

Whisk yolks into milk mixture and stir for 1 minute, then stir in gelatine and continue to stir until it has all dissolved.

Remove from heat and whisk in butter and remaining orange zest.

Cover surface closely with plastic cling film.

Allow pastry cream to cool to room temperature.

To prepare the peaches:

Remove peel and pit from 6 peaches, then cut flesh into 5mm-thick (about ¼-inch) slices.

Half-fill a large bowl with water and ice cubes.

Pour sugar and 60ml (¼ cup) water into a small saucepan and simmer over medium heat until it becomes a golden caramel.

Dip base of pan into the bowl of ice-water to stop the cooking.

Cool for 2 minutes, then drizzle base of tart with enough caramel to lightly coat it.

Add juice from half the remaining peach to caramel in saucepan. Return to stove-top and stir over low heat until smooth.

To assemble the tart:

Fill tart shell with pastry cream and top with peach slices. Drizzle syrup over the top, then put tart in freezer for 10 minutes to firm and harden the caramel before serving.

EASY RED-FLESHED PEACH TART

A French recipe. Vegetarian. Serves 6 people.

Ingredients
- 7 red-fleshed peaches
- 1 sheet ready-rolled frozen shortcrust pastry, defrosted
- 3 large free-range eggs
- ⅓ cup sugar
- 2½ cups milk

Instructions

Preheat oven to 200 °C (400 °F).

Line a pie dish with the pastry and prick the base with a fork.

Peel peaches [see page 63], and cut them in half and then into thick slices.

Arrange them decoratively on the pastry.

Beat egg with milk and add sugar.

Pour this mixture on the tart.

Put in the oven and bake about 25 minutes until the egg custard is set.

Serve warm or cold, with ice cream or cream.

RED-FLESHED PEACH TART WITH MASCARPONE MOUSSE

A French recipe; *'Tarte aux pêches de vigne et à la mousse de mascarpone'*. Vegetarian. Serves 8 people.

Ingredients
- 250g (9oz) flour
- 30 + 30 + 50 g (1 + 1 + 2 oz) sugar
- 100g (3½ oz) soft butter
- 1 egg yolk
- 5 red-fleshed peaches
- 250g (9oz) mascarpone
- 2 large eggs, separated

Instructions
Preheat oven to 220 °C (430 °F).

For the pastry shell: Combine flour with 30g (1oz) of sugar, the egg yolk and butter.

Add a small amount of water and mix to form a soft paste.

Roll out the dough and press it into an ovenproof tart pan or pie plate.

Cover the dough with baking paper and sprinkle it with dried beans to weigh it down. Bake for about 30 minutes until the dough is golden. Allow pastry to cool.

For the mascarpone mousse: In a bowl, whisk the other 2 egg yolks and 30g (1oz) sugar until mixture is frothy.

Add the mascarpone, continuing to whisk.

Beat the 2 egg whites. Fold them into the mascarpone mixture.

To prepare the fruit:
Peel the peaches [see page 63] and slice them.

Heat 100 ml (6 ½ tablespoons) of water with 50g (2oz) of sugar. When the mixture comes to a boil, add the peach slices. Let them poach for one minute, then drain in a sieve and cool them by rinsing under cold water for 30 seconds

To assemble the tart:
Pour the mascarpone mousse into the pastry shell, and top with peach slices. Refrigerate until ready to serve.

RED-FLESHED PEACH TARTE TATIN

Vegetarian

A 'Tarte Tatin' is an upside-down fruit pastry in which the fruits are caramelized in butter and sugar before the tart is baked. Vegetarian. Serves 6 people.

Ingredients
- 1.5kg (3 - 4 pounds) red-fleshed peaches
- 250g (9oz) flaky or puff pastry
- 150g (5 oz) castor sugar,
- 60g (2oz) butter

Special equipment
- A baking pan with a removable base.

Instructions
Preheat oven to 180 °C (350 °F).

In a saucepan, caramelise the sugar with the butter. Pour the caramel into the baking tin.

Cut the peaches in half, scoop out the stones and slice them into wedges of a uniform thickness.

Arrange the wedges closely together on top of the caramel in the baking pan.

Place in preheated oven and bake for about 20 minutes.

Remove pan from oven and allow to cool.

Roll out the flaky or puff pastry to a thickness of 2.5 cm.

Cut out a circle of pastry the same size as the pan and completely cover the fruit with it. Trim edges as needed.

Return tart to the oven and bake for 30 minutes until the pastry is well cooked.

Take tart out of the oven and leave it to cool for 2 hours.

Remove the tart from the tin by placing an oven-proof dish over the flaky/puff pastry and turning it over.

Place the tart back in the oven for a few minutes and serve slightly warm.

ROSY RED-FLESHED PEACH TART

Vegetarian, gluten-free

Ingredients

For shortcrust pastry:
- 113g (4oz) salted butter, softened
- 68g (2½ oz) sugar
- 1 egg yolk from large egg, at room temperature
- 270g (9½ oz) gluten-free self-raising flour[50]

For the filling and topping:
- 226 g (8 oz) low fat cream cheese at room temperature
- zest of 1 lemon
- 100g (3½ oz) icing sugar (confectioners' sugar)
- 1 teaspoon vanilla extract
- pinch of salt
- a dash of nutmeg (or cinnamon, mace or allspice)
- 1 ripe peach
- 1 cup apple jelly (or quince, currant or grape jelly)
- freshly-picked rosebuds for garnish (optional)

Instructions

Preheat oven to 180 °C (350 °F).

For shortcrust pastry:

Using a stand mixer equipped with a paddle attachment, mix the butter, sugar, and salt at low speed until creamy.

Add the egg yolk and mix until it is just absorbed.

Slowly add in the flour mixture and mix until evenly blended into a dough.

Roll out the dough between two sheets of parchment or wax paper until it is very thin, about ½cm (¼ inch). Chill on a biscuit tray (cookie sheet) in the fridge for at least 30 minutes, until it firms up enough to work with it, but is not hard and rigid.

---

50   To make self-raising flour, add 2 teaspoons of baking powder and 1/2 teaspoon salt to every cup of plain (all-purpose) flour.

Remove dough from refrigerator, take off the top sheet of paper and carefully turn it over into a tart pan or pie dish.

Working from the middle, push the dough with your fingers to the bottom of the pan while removing the other layer of paper. Ensure that you push the dough right into the corners.

Using a butter knife, cut off the excess dough from the edges of the pan.

Prick the dough with a fork, then bake in preheated oven for about 20 minutes, until brown.

Remove and allow to cool.

For the filling:
In a medium bowl, mix together the softened cream cheese and the lemon zest.

Add the sugar, salt, mace (or other spice) and vanilla. Stir until well-combined.

Pour the cream cheese mixture into the cooled pie shell, then flatten and smooth the surface with a soft spatula.

For the topping:
Slice a peach into very thin pieces. Arrange them decoratively on top of the tart filling.

Heat fruit jelly until it becomes runny.

Remove from heat.

Using a pastry brush, gently smooth fruit jelly evenly over the peaches and filling.

If using rosebuds, dip them into melted fruit jelly and place on tart.

Chill tart for a minimum of 1 hour before serving.

Store in refrigerator.

Red-fleshed Peach Cakes and Slices

RED-FLESHED PEACH SLICE

Vegetarian

Ingredients

Base:
- 1 cup wheat bran
- ½ cup coconut
- ½ cup whole almonds
- 60g (4 tablespoons) ricotta

Filling:
- 800g red-fleshed peaches stewed in water and drained (enjoy the juice as a chilled drink)
- 2 teaspoons pomegranate molasses
- 1 teaspoon cardamom
- ½ cup raw sugar

Topping:
- 250g (9 oz) flaked almonds

Instructions

Preheat oven to 190 °C (375 °F).

Combine all base ingredients thoroughly in a food processor then press into a shallow, foil-lined cake pan.

Mix together filling ingredients and spread evenly over base.

Liberally sprinkle the flaked almonds on top.

Bake for ¾ hour or until almonds are golden brown.

Cool, refrigerate.

Cut into slices and serve.

RED-FLESHED PEACH AND APPLE UPSIDE-DOWN CAKE

Vegetarian

Ingredients
500g (1.1 pounds) red-fleshed peaches washed
1200g (2 ½ pounds) apples, peeled and coarsely grated
5 tablespoons sugar
1 teaspoon ground cinnamon
¼ teaspoon cardamom powder
1 pinch nutmeg
5 tablespoons water
3 tablespoons cornstarch
3 eggs
3 tablespoons sugar
4 tablespoons flour
½ tablespoon baking powder
butter, to grease the cake pan

Instructions
Pre-heat the oven to 180 °C (350 °F).

Grease the cake pan with butter, line it with baking paper and grease the paper again.

Cut peaches into slices and arrange them decoratively in the base of the pan.

Mix the apples, sugar and the spices together in a saucepan and cook over a medium heat until fruit begins to soften. Remove saucepan from stove top.

Mix the water with the cornstarch and stir it into the apple mixture. Return saucepan to stove top and cook again, stirring, until mixture thickens.

Pour this hot mixture carefully onto the peaches without moving the slices.

In the bowl of an electric mixer beat eggs with sugar until soft and creamy. Add the flour and the baking powder and mix to form a soft dough. Turn the dough out of the bowl and place it on top of the fruit, covering it.

Put into oven and bake for 35 minutes.

Remove from oven. After cake cools down, cover it with a plate, turn it upside down and remove the baking paper.

Red-fleshed Peach Drinks

SPARKLING RED-FLESHED PEACH SANGRIA

Vegan, gluten-free.

Ingredients
- 1 cup water
- ⅓ cup tightly-packed brown sugar
- 3 large, ripe peaches
- 2 ½ cups dry white wine
- ⅓ cup orange liqueur (such as Grand Marnier)
- 750 ml chilled cava or other sparkling white wine

To decorate: blueberries, mint leaves and peach slices

Special equipment
cheesecloth or muslin

Instructions

Combine water and brown sugar in a microwave-safe bowl.

Microwave on 'high' for 3½ minutes. Stir to dissolve sugar, then allow the sugar syrup to cool.

Peel and pit peaches [see "How to Peel Peaches" on page 63].

Place sugar syrup and peeled peaches in a food processor and process until smooth.

Pour peach mixture into a jug and stir in orange liqueur.

Chill for at least 4 hours.

Strain mixture through a cheesecloth or muslin-lined sieve. Squeeze cloth to extract juices.

Discard solids.

Stir in white wine and decorate before serving.

RED_FLESHED PEACH AND BLUEBERRY SPARKLER

Vegan, gluten-free. Serves 3 people.

Ingredients
1 red-fleshed peach, chilled
a large handful of blueberries, chilled
4 shots of silver rum (a spicy white/light rum)
champagne
ice cubes

Instructions

Chill stemless wine glasses in the freezer.

Puree one peach and an overflowing handful of blueberries, reserving 3 blueberries

Add four shots of silver rum. Blend.

Fill the wine glasses with ice cubes, add the mixture and top with champagne.

Garnish with 3 blueberries.

Red-fleshed Peach Preserves, Conserves and Pickles

BRANDIED RED-FLESHED PEACHES
Vegan, gluten-free

Ingredients
- 3 to 4 kg (6 to 8 pounds) firm red-fleshed peaches
- 1 ¼ cups sugar
- 1 cup water
- 1 ½ cups brandy
- lemon juice

Special equipment
Sturdy, sealable glass preserving jars, pre-heated
A hot-water bath for sterilising preserves.

Instructions
Combine sugar and water in a saucepan, bring to a boil, and cook until sugar dissolves. Set aside but keep the syrup hot.

Dip the peaches into boiling water for ½ to 1 minute to loosen skins, then immediately plunge them into a bowl of ice-water [see "How to Peel Peaches" on page 63].

Peel and halve the peaches and remove the stones.

Dip fruit into lemon juice to prevent it from discolouring.

Drain the peaches and pack firmly into hot jars, cavity sides facing down.

Pour in about ¼ cup hot sugar syrup. Next, pour 3 to 4 tablespoons of brandy over fruit. Add enough syrup to fill the jars to within 1cm (½") of rim.

Put on the lid and make sure it is tightly sealed.

Process in a simmering hot-water bath for 25 minutes, according to manufacturer's instructions.

RED-FLESHED PEACH SPICED PICKLES

An attractive sweet and sour pickle. Vegan and gluten-free.

Ingredients
- 2kg (4.4 pounds) red-fleshed peaches, halved, stones removed.
- 1kg (2.2 pounds) sugar
- 600ml (1 pint) apple cider vinegar
- 12 black peppercorns
- 4 cardamom pods
- 2 star anise
- 2 cinnamon sticks
- 1 teaspoon whole cloves

Equipment
Four clean glass jars of around 500ml (approx. 1 pint) capacity, with lids that seal tightly.

Instructions
In a saucepan, bring apple cider vinegar and spices to a boil.

Add the sugar and stir the mixture over a low heat until sugar is fully dissolved.

Bring the syrup back to a rolling boil, add the peach halves and simmer for 4-5 minutes until just soft. Do not overcook. Prick peaches with a sharp knife to test whether they are soft.

To sterilise the jars:
- Heat your oven to no more than 275 °F/130 °C/Gas 1.
- Place a double layer of newspaper on each oven shelf.
- Arrange the jars on the shelf ensuring that they do not touch each other.
- Close the oven door and leave the jars in the hot oven for for a minimum of 20 minutes.

- Put on a pair of well-insulated oven mitts then take each jar from the oven and place onto a heatproof mat.

Transfer hot peaches to hot sterilized jars. (Never place hot food in cold glass jars or cold food in hot glass jars or the glass may shatter.)

Boil the syrup for a few more minutes to reduce it slightly.

Pour syrup over the peaches, distributing the spices evenly between the jars, until the peaches are completely covered.

Seal the jars, label them and leave them in a cool, dark place for two months or more to allow the flavours to develop.

RED-FLESHED PEACH CHUTNEY

Vegan and gluten-free.

Ingredients
- 20 red-fleshed peaches
- 2 onions, peeled and chopped
- 250 ml (1 cup) malt vinegar
- 200g (7 oz) brown sugar
- 10 whole cloves
- 1 cinnamon stick
- ½ teaspoon chilli powder
- ½ teaspoon ground cinnamon
- ½ teaspoon salt

Instructions

Heat the oven to 240° C (460° F).

Peel the peaches [see "How to Peel Peaches" on page 63].

Halve the peaches and remove the stones, then arrange them in a single layer in a large baking dish or roasting pan.

Sprinkle the chopped onions on the peaches.

Roast the peaches and onions in the oven for 30-40 minutes, until all juice in the pan has evaporated and the peaches have shrunk a little.

Weigh the roasted peach/onion mix. There should be about 1 kg.

Put the peach/onion mix into a large, heavy-based saucepan.

For every 900-1100 g (2 - 2½ pounds) of peach/onion mixture, add the sugar, vinegar and spices in the amounts shown above. Adjust the amounts up or down if the peach/onion mix weight is heavier or lighter.

Bring the saucepan to the boil, then reduce the heat and simmer slowly, stirring often, until the mixture thickens. This should take around 30 minutes.

Remove the cinnamon stick and bottle the chutney in hot, sterilised glass jars.

Seal immediately.

RICH RED-FLESHED PEACH CHUTNEY WITH RAISINS, NUTS AND GINGER

Vegan and gluten-free.

Ingredients
- 1.5kg (3½ pounds) firm peaches, diced
- 1 ½ cups seedless golden raisins
- 1 large onion, minced
- 1 yellow pepper, diced
- 2 hot red peppers, diced (remove seeds)
- ½ cup crystallized ginger, chopped
- ½ cup pecans, very coarsely chopped
- 2 cups cider vinegar
- 3 cups sugar
- 1 teaspoon cinnamon
- ¼ teaspoon ground cloves
- ¼ teaspoon mace or nutmeg

Special equipment
Sturdy, sealable glass preserving jars, pre-heated
A hot-water bath for sterilising preserves.

Instructions
Mix all the ingredients except the nuts in a heavy saucepan.
Bring to a boil, stirring constantly.
Turn down heat and simmer until the chutney is thick enough to form a mound on a spoon.
Remove the chutney from the heat and stir in the nuts.
Ladle chutney into hot preserving jars leaving 1cm (½-inch) headspace.
Affix the lid and seal tightly.
Process in a simmering hot water-bath for 10 minutes.

FRENCH RED-FLESHED PEACH JAM

A delicious conserve — not too sweet, with an incomparable flavour and a beautiful, deep red colour. Vegan and gluten-free.

Ingredients
- 1 kg peaches
- 600g sugar

Instructions
Peel peaches [see "How to Peel Peaches" on page 63] and dice them. Put them in a bowl and pour the sugar over. Mix well.

Allow to stand for about 1 hour.

Pour the sugar and peaches into a saucepan. Heat gently for about 20 minutes, then turn off the heat and skim off the foam that may be floating on top.

Pour jam into clean, pre-warmed glass jars. Allow it to cool.

RED-FLESHED PEACH CONFITURE

A French recipe: '*Confiture de pêches de vigne*'
Vegan and gluten-free. Makes 5 jars.

Ingredients
- 2 kg (4.4 pounds) red-fleshed peaches
- 500g (17oz) sugar
- 40g (2 ½ tablespoons) pectin powder
- 2 vanilla pods
- ½ teaspoon cinnamon
- lemon juice

Special equipment
Dry, sterilised glass jars[51] with tight-fitting lids

Instructions
Peel the peaches [see "How to Peel Peaches" on page 63]. Remove the stone, cut peaches into small pieces.

Place into a saucepan the peaches, sugar, pectin, lemon juice, vanilla beans and cinnamon.

Cook for 10 minutes over a medium heat, stirring occasionally. Check the sugar content by tasting the mixture, and modify according to your taste. Do not overcook or the fruit will become pureed.

The confiture is cooked when it is soft but there are still some whole pieces of fruit.

Slowly pour into heated jars and seal.

Label your jars.

---

[51] *Sterilise jars by putting them in cold water, bringing them to a boil and boiling for a few minutes. Place them on a clean cloth to air-dry.*

RED-FLESHED PEACH JAM WITH BERRIES AND RHUBARB

Vegan and gluten-free.

Ingredients
- 1100g (2½ pounds) fresh or frozen red-fleshed peaches, halved, stones removed
- 972g (2.1 pounds) sugar
- 240g (1 cup) water
- half a vanilla pod
- 3 x red rhubarb stalks, finely chopped
- ½ cup frozen blackcurrants
- juice of 1½ lemons

Instructions

Mix together the sugar, water and vanilla and boil rapidly for 5 minutes or until it turns into a thick syrup.

Add the peaches and rhubarb and bring back to the boil.

Boil rapidly for 40 minutes, stirring often.

Add the currants and lemon juice and boil for another 5 minutes or until a little of the jam spooned onto a saucer wrinkles on the surface after it has cooled.

Remove jam from heat.

Push jam through a sieve to give it a smooth texture — or you can blend it with a stick blender.

Pour hot jam into hot sterilized jars and immediately put the lids on. Allowing to cool with lids on creates a protective vacuum seal.

Red-fleshed Peach Savoury Dishes and Salads

SALMON WITH RED-FLESHED PEACH SALSA

Gluten-free.

Ingredients

Salsa:
- 3 large red-fleshed peaches, diced
- 1 bunch fresh coriander (cilantro), washed and chopped
- 1-2 fresh jalapeño[52], diced
- 2 cloves garlic, minced
- juice of 1 lime
- salt and pepper to taste

Salmon:
- 4-6oz salmon fillets
- 2 tbs olive oil
- 2 cloves garlic, minced

Instructions

For the salsa:

In large bowl mix together peaches, cilantro, jalapeños, garlic and lime juice.

Taste and add salt and pepper as needed. Set aside.

Preheat large heavy frypan (skillet) over medium heat for 3-5 minutes.

Coat salmon with olive oil and minced garlic.

Place fish in frypan and increase heat to high.

Cook for 3 minutes.

Flip the fish over and fry for an additional 5 minutes or until salmon is flaky and firm.

Serve hot, drizzled with salsa, accompanied by green salad.

---

52   *To adjust the heat of the salsa, feel free to remove seeds and choose either one or two jalapeños for this mix. I like the heat of the jalapeño and chose to make my salsa with two jalapeños, seeds included.*

SCALLOPS TARTARE WITH RED-FLESHED PEACHES

Gluten-free. Serves 4 people.

Ingredients

- One quantity of red-fleshed peach compote made according to the recipe for 'Red-fleshed Peach Compote' on page 73 and then allowed to cool.
- 6 red-fleshed peaches, washed
- 12 shelled scallops without coral
- juice of half a lemon
- 1 tablespoon of olive oil
- a few sprigs of thyme
- pepper

Instructions

In a bowl, mix together the olive oil, lemon juice, thyme and a crunch of pepper.

Cut the scallops into small pieces and marinade them for 1 hour in the olive oil mixture.

Remove the stones from the 6 red-fleshed peaches and cut into thin slices.

Fill a ramekin to three-quarters of its height with the marinated scallops tartare, firming them down.

Fill to the top with the peach compote.

Turn out this preparation on a dinner plate placing the peach slices decoratively all around it, and garnish it with a final slice.

Serve chilled.

RED-FLESHED PEACH AND CHÈVRE SALAD

Vegetarian and gluten-free. Serves 2 people.

Ingredients
- 2 ripe red-fleshed peaches, sliced, stones removed.
- 2 medium sized tomatoes
- 4 tablespoons olive or grapeseed oil
- ½ teaspoon salt
- 2 tablespoons lemon juice
- 3 teaspoons balsamic vinegar
- ½ teaspoon fresh or dried thyme
- chèvre (goat cheese)

Instructions

For the vinaigrette:

In a jar or medium sized bowl, combine oil, salt, lemon juice, balsamic vinegar and thyme. Mix well.

On plate, assemble sliced peaches, tomatoes, and crumbled chèvre.

Drizzle salad with vinaigrette just before serving.

RED-FLESHED PEACH CARPACCIO SALAD

A Carpaccio is an Italian *hors d'oeuvre* consisting of thin slices of raw beef or fish served with a sauce. This variation uses fruit instead. Vegetarian, gluten-free.

Ingredients
- 2 large red-fleshed peaches
- ¼ cup walnuts, chopped into small pieces
- 1-2 tablespoons honey
- 2 large handfuls of rocket (arugula), washed
- 1 shallot, finely sliced (or substitute spring onions or the white part of a leek)
- 3-4 tablespoons of soft goat cheese, crumbled
- balsamic or white wine vinegar
- salt
- pepper

Instructions

Preheat oven to 150 °C (300 °F).

With a sharp knife or a mandolin, cut the peaches into very thin slices.

Spread the walnuts on a baking tray (sheet) and roast them in the oven for about 10 minutes, stirring occasionally. When they are roasted, remove them from the oven and allow them to cool slightly. Put them in a small bowl, add the honey and mix, then set aside.

Arrange peach slices on a small plate so that they overlap slightly, in an attractive spiral.

Sprinkle the shallots, goat cheese and walnuts on top on the peach slices. Top with the rocket (arugula).

Drizzle lightly with vinegar and season with salt and pepper.

RED-FLESHED PEACH SALAD WITH BLUE CHEESE

Vegetarian, gluten-free. Serves 4 people.

Ingredients
2 firm red-fleshed peaches sliced, then lightly grilled
1 lettuce
¼ cup slivered almonds, toasted
30 - 60g (1 - 2 oz) crumbled blue cheese

Peach Vinaigrette
¼ cup apple cider vinegar (or white vinegar)
1 tablespoon peach jam [see page 100] [53]
1 teaspoon Dijon mustard
½ cup lemon-infused or plain olive oil
sea salt
freshly ground black pepper

Instructions
For the vinaigrette: Put vinegar, mustard, olive oil and peach jam into a jar. Screw the lid on tightly and shake jar until all ingredients are well mixed.

For the salad: Wash and dry lettuce. Tear into bite-size pieces. Mix lettuce with crumbled blue cheese and most of the toasted almonds, saving some cheese and nuts for garnish.

Apply a small amount of dressing to lettuce, just enough to wet the leaves, and toss gently with your hands.

Season with salt and pepper.

Arrange fried peaches on top of salad with a little crumbled blue cheese and toasted almonds. Drizzle vinaigrette over the top and garnish with cheese and nuts immediately before serving.

---

53   *If you don't have any peach jam, substitute fresh peach slices that have been grilled.*

RED-FLESHED PEACH LIQUEUR

Crème de Pêche de Vigne

Crème de pêche, or peach liqueur as it is known in English, is obviously made from peaches. It is one of the more unusual fruit liqueurs, though many of the big liqueur companies produce at least one form of it. Crème de pêche is often quite sweet, with an intense flavour of ripe peaches and occasionally with a subtle, fragrant flavour similar to almonds, imparted whenever peach stones are included in the production process.

As well as standard crème de pêche liqueur there is also crème de pêche de vigne, which is made from the vineyard peach. The Pêche de Vigne has a more intense, aromatic flavour than other peaches. The resulting liqueur tends to me more complex than plain crème de pêche, though it also it also has a kind of nuanced delicateness.

Cream or Crème?

A crème liqueur is distinct from a cream liqueur. A crème liqueur contains sufficient sugar to give it a full-bodied, syrupy consistency. Unlike cream liqueurs, crème liqueurs contain no cream. 'Crème' in this sense refers to the viscosity. Examples of crème liqueurs include crème de cacao (chocolate), crème de menthe (mint), crème de mûre (blackberry), crème de cassis (blackcurrant), crème de griottes sauvages (wild morello cherry), crème de myrtilles (blueberry), crème de framboise (raspberry), and crème de pêche (peach).

Custom-made Crème de Pêche de Vigne

Late one summer we were blessed with a glut of pêches de vigne at our farm on the Mornington Peninsula. Hundreds of glorious, dusky-red fruits weighed down the branches of the trees in our orchard. The local greengrocers were buying boxfuls from us, and couldn't get enough to satisfy their customers. My own freezer was full of them and the fruit dryer was running day and night.

Even while we basked in this abundance we knew it would be all too brief. The season would be over in a few short weeks and I longed for a way to preserve the goodness, colour, flavour and beauty of our pêches de vigne in ways other than the usual jams and jellies. It occurred to me to experiment with making liqueurs.

I had never made liqueurs before but once the idea had struck me it excited me so much that I dived in and researched the topic thoroughly. Hunting avidly through recipe books while armed with high school French I immersed myself in vintage, tried-and-true recipes from such venerable tomes as —

- The Whole Art of Making British Wines, Cordials, and Liqueurs, in The Greatest Perfection; as also, Strong and Cordial Waters: to Which is Added a Collection of Valuable Recipes for Brewing Fine and Strong Welsh Ales, and Miscellaneous Articles Connected with the Practice. By James Robinson, published 1848.
- Nouveau Manuel Complet de la Fabrication des Vis de Fruits du Cidre, du Poiré, des Boissons Rafraich-issantes, des Bières Économiques et de Ménage, des Vins de Grains, des Hydromels, de Boissons Diverses, et D'imiter les Vins de Liqueur Français et Étrangers.[54] Consolidated And Increased Considerably, by Mr F. M Alepetre. Librairie Encyclopédique De Roret, published 1854.
- Nouveau Manuel Complet du Distillateur Liquoriste: Contenant l'art de Fabriquer les Sirops, les Esprits Parfumés, les Huiles Essentielles, les Eaux Distillées,

---

54   *This roughly translates as 'Complete New Manual of the Manufacturing of Fruit Cider, Perry, Refreshing Beverages, Beer for Business and Household, Grain Spirits, Meads, Various Drinks and Imitations of the French and foreign liqueurs'..*

les Ratafias et Les Hypocras, Renfermant Toutes les Recettes.[55] By Nicolas Lebeaud, published 1918.
- Old-Time Recipes for Home Made Wines, Cordials and Liqueurs from Fruits, Flowers, Vegetables, and Shrubs. By Helen S. Wright (Helen Saunders), published 1922.
- Chemistry and Technology: Wines and Liquors. By Karl M. Herstein and Thomas C. Gregory, published 1935.

Using our red-fleshed peaches and a few other exquisite botanicals from plants organically grown at our farm I began to experiment, testing a wide variety of methods and conducting innumerable (pleasurable) taste tests.

Soon the pantry shelves were lined with boxes of meticulously-labelled Mason jars. I kept careful records of every step in the process of devising my recipe – how the peaches were prepared, how long each stage lasted, the number and kind of added botanicals, the exact proportions of all ingredients, how the filtering was carried out etc.

And at long last — what a result! The finished crème de pêche de vigne liqueur exceeded my expectations. The colour was a dark, limpid red; the colour of sunset glowing through ruby glass. When poured, it flowed with a slow, velvety smoothness. At the first sip a rosy haze seemed to rise up and suffuse me. How could I define this complex flavour? My fondness for words came to the rescue. I could only describe it as a blend of blackcurrants; the dark chocolate and cassis of Shiraz wine, the richness of purple plums and a hint of the fragrance of musk roses.

Years ago when I planted our first red-fleshed peach tree, little could I guess that the growing of these interesting fruits in the local terroir would lead to the writing of a book and the creation of a truly remarkable liqueur!

C. Thornton

---

55 *'Complete New Manual of the Liqueur Distiller: Containing the Art of Making Syrups, Perfumed Spirits, Essential Oils, Distilled Water, and Ratafias and Mead; Containing All Recipes.'*

Jean de la Quintinye
Directeur de tous les Jardins Fruitiers et Potagers du Roy

Index

A

about peaches 11
about rare and heritage fruit 1
Africa xii
Akaroa 33
anthocyanins 20
anti-inflammatory 21

B

bioflavonoids 21
Blackboy Peach xii, 31
Black Plumb Peach of Georgia 44
Bordeaux 26
brugnon 25

C

California 19
Cape of Good Hope 37
cardiovascular disease 22
Charente 35
Cherokee 41
chill hours 18
China xii
climate xi
clingstone 13
crème de pêche de vigne 108

D

diabetes. 22
Dutch 37
Dutch East India Company 37

E

espalier 23
eyesight 22

F

flavonoids 20
flesh colour 13
flesh texture 14
France 11
freestone 13

G

grafting 13

H

health benefits of red-fleshed
     peaches 20
heritage apples xi
heritage fruit xi
history 11
how to peel peaches 63
Huguenots 38

I

Île de France 23
Indian Blood 13
Indian Blood Cling 39
Indian Blood Free 39

J

Japan 21
Jefferson 43

K

King Louis Philippe 33

L

Langlois 33
latitude xi

M

Mediterranean 19
Montreuil-aux-Pêches 23
Mornington Peninsula xiii

N

names of red-fleshed Pêches de Vigne 28
Native Americans 40
nectarines 15, 25
Nectavigne 25
New Zealand xii

O

obesity 22

P

pavie 25
peach blossoms 15
pêche de vigne xii, 26
Pêche de Vigne Festival 30
pêche sanguine 13
pêche sanguine vineuse 13
pêche vraie 25
Persia 11
persica 12
'Pit' or 'Stone' 13
pleach 13
pleurisy 22
Prunus 12

Q

Quintinie 24

R

red-fleshed peaches xii
red-fleshed peaches in Africa 37
red-fleshed peaches in Australia 49
red-fleshed peaches in France 23
Red-Fleshed Peaches in New Zealand 31
red-fleshed peaches In The USA 39
red-fleshed peaches on the Mornington Peninsula 59
red-fleshed peach liqueur 108
red-fleshed peach recipes 63
Red Gem 19
Romans 11

S

scientific classification 12
self-fertile 16
semiclingstone 13
semifreestone 13
Shangdong Province 19
Soucieu-en-Jarrest 19
Sun King 24

T

Tasmania-France Connection 51
terroir 17
the Americas 12
true-to-type 16

U

USA 13
uses 14

V

Versailles 24
volcanic soils xi

W

white-fleshed 20

Y

yellow-fleshed 20

Some Heritage Fruit Groups in Australia

Werribee Park Heritage Orchard, situated near Melbourne, Australia, is a beautiful antique orchard dating from the 1870s, on the grounds of the old mansion by the Werribee River. It was renowned for its peaches, grapes, apples, quinces, pears, a variety of plums and several other fruits, as well as walnuts and olives. Recently this historic treasure has been rediscovered. Volunteers are replanting and tending the orchard.

www.werribeeparkheritageorchard.org.au

The Heritage Fruits Society is also based in Melbourne, Australia. Their aim is to conserve heritage fruit varieties on private and public land. They enable and encourage society members to research this wide range of varieties and to inform the public on the benefits of heritage fruits for health, sustainability and biodiversity.

www.heritagefruitssociety.org.au

The Heritage and Rare Fruit Network's purpose is to provide a forum for sharing information on all varieties of fruit and less common useful plants, to link up people with an interest in growing unusual fruit, and to support sharing of propagation material through grafting days and any other means.

heritageandrarefruits.weebly.com

The Rare Fruit Society of South Australia is an amateur organisation of fruit tree growers who preserve heritage varieties, explore climate limitations and study propagation, pruning and grafting techniques.

www.rarefruit-sa.org.au

Rare and Heritage Fruit series

www.ingramcontent.com/pod-product-compliance
Lightning Source LLC
LaVergne TN
LVHW051501070426
835507LV00022B/2868